Delmar's Test Preparation Series

Medium/Heavy Truck Test

Gasoline Engines (Test T1)

**Technical Advisor
John F. Kershaw**

Africa • Australia • Canada • Denmark • Japan • Mexico • New Zealand • Philippines
Puerto Rico • Singapore • Spain • United Kingdom • United States

NOTICE TO THE READER

Publisher does not warrant or guarantee any of the products described herein or perform any independent analysis in connection with any of the product information contained herein. Publisher does not assume, and expressly disclaims, any obligation to obtain and include information other than that provided to it by the manufacturer.

The reader is expressly warned to consider and adopt all safety precautions that might be indicated by the activities herein and to avoid all potential hazards. By following the instructions contained herein, the reader willingly assumes all risks in connection with such instructions.

The Publisher makes no representation or warranties of any kind, including but not limited to, the warranties of fitness for particular purpose or merchantability, nor are any such representations implied with respect to the material set forth herein, and the publisher takes no responsibility with respect to such material. The publisher shall not be liable for any special, consequential, or exemplary damages resulting, in whole or part, from the readers' use of, or reliance upon, this material.

Delmar Staff:
Business Unit Director: Alar Elken
Product Development Manager: Jack Erjavec
Executive Marketing Manager: Maura Theriault
Channel Manager: Mona Caron
Executive Production Manager: Mary Ellen Black
Fast Cycle Production Editor: Betsy Hough
Cover Design: Paul Roseneck
Cover Image: © 1998 Corbis Corp.

COPYRIGHT © 2000
Delmar is a division of Thomson Learning. The Thomson Learning logo is a registered trademark used herein under license.

Printed in Canada
3 4 5 6 7 8 9 10 XXX 05 04 03 02 01 00

For more information, contact Delmar, 3 Columbia Circle, PO Box 15015, Albany, NY 12212-0515; or find us on the World Wide Web at http://www.delmar.com.

All rights reserved Thomson Learning 2000. The text of this publication, or any part thereof, may not be reproduced or transmitted in any form or by any means, electronics or mechanical, including photocopying, recording, storage in an information retrieval system, or otherwise, without prior permission of the publisher.

You can request permission to use material from this text through the following phone and fax numbers. Phone: 1-800-730-2214; Fax 1-800-730-2215; or visit our Web site at http://www.thomsonrights.com.

ISBN 0-7668-0559-X

Contents

Preface vi

Section 1 The History of ASE

History 1
   NIASE 1
   The Series and Individual Tests 2
   A Brief Chronology 2
   By the Numbers 3
   ASE 4

Section 2 Take and Pass Every ASE Test

ASE Testing 7
   Who Writes the Questions? 7
   Testing 8
   Be Test-Wise 8
   Before the Test 8
   Objective Tests 9
   Taking an Objective Test 9
   During the Test 10
   Review Your Answers 10
   Don't Be Distracted 10
   Use Your Time Wisely 11
   Don't Cheat 11
   Be Confident 11
   Anxiety and Fear 12
   Getting Rid of Fear 12
   Effective Study 13
   Make Study Definite 13
   The Urge to Learn 14
   Concentrate 14
      Get Sufficient Sleep 15
      Arrange Your Area 15
      Don't Daydream 15

Study Regularly 16
Keep a Record 17
Scoring the ASE Test 17
Understand the Test Results 18
    Gasoline Engines (Test T1) 18
"Average" 19
So, How Did You Do? 19

Section 3 Are You Sure You're Ready for Test T1?

Pretest 21
    Answers to the Test Questions for the Pretest 23
    Explanations to the Answers for the Pretest 23
Types of Questions 26
    Multiple-Choice Questions 26
    EXCEPT Questions 26
    Technician A, Technician B Questions 27
    Questions with a Figure 28
    Most-Likely Questions 28
    LEAST-Likely Questions 29
    Summary 29
Testing Time Length 30
    Monitor Your Progress 30
    Registration 30

Section 4 An Overview of the System

Gasoline Engines (Test T1) 31
    Task List and Overview 32
    A. General Engine Diagnosis (15 Questions) 32
    B. Cylinder Head and Valve Train
       Diagnosis and Repair (8 Questions) 35
    C. Engine Block Diagnosis and Repair (8 Questions) ... 38
    D. Lubrication and Cooling Systems
       Diagnosis and Repair (8 Questions) 42
    E. Ignition System Diagnosis and Repair (11 Questions) ... 44
    F. Fuel and Exhaust Systems
       Diagnosis and Repair (10 Questions) 47

G. Battery and Starting Systems
   Diagnosis and Repair (7 Questions) 50
H. Emissions Control Systems
   Diagnosis and Repair (7 Questions) 51
I. Computerized Engine Controls
   Diagnosis and Repair (6 Questions) 53

Section 5 Sample Test for Practice

Sample Test 57

Section 6 Additional Test Questions for Practice

Additional Test Questions 73

Section 7 Appendices

Answers to the Test Questions for the Sample Test Section 5 97
Explanations to the Answers for the Sample Test Section 5 98
Answers to the Test Questions for the Additional Test Questions
    Section 6 110
Explanations to the Answers for the Additional Test Questions
    Section 6 111
Glossary 127

Preface

This book is just one of a comprehensive series designed to prepare technicians to take and pass every ASE test. Delmar's series covers all of the Automotive tests A1 through A8 as well as Advanced Engine Performance L1 and Parts Specialist P2. The series also covers the five Collision Repair tests and the eight Medium/Heavy Duty truck test.

Before any book in this series was written, Delmar staff met with and surveyed technicians and shop owners who have taken ASE tests and have used other preparatory materials. We found that they wanted, first and foremost, *lots* of practice tests and questions. Each book in our series contains a general knowledge pretest, a sample test, and additional practice questions. You will be hard-pressed to find a test prep book with more questions for you to practice with. We have worked hard to ensure that these questions match the ASE style in types of questions, quantities, and level of difficulty.

Technicians also told us that they wanted to understand the ASE test and to have practical information about what they should expect. We have provided that as well, including a history of ASE and a section devoted to helping the technician "Take and Pass Every ASE Test" with case studies, test-taking strategies, and test formats.

Finally, techs wanted refresher information and reference. Each of our books includes an overview section that is referenced to the task list. The complete task lists for each test appear in each book for the user's reference. There is also a complete glossary of terms for each booklet.

So whether you're looking for a sample test and a few extra questions to practice with or a complete introduction to ASE testing, with support for preparing thoroughly, this book series is an excellent answer.

We hope you benefit from this book and that you pass every ASE test you take!

Your comments, both positive and negative, are certainly encouraged! Please contact us at:

Automotive Editor
Delmar Publishers
3 Columbia Circle
Box 15015
Albany, NY 12212-5015

1 The History of ASE

History

Originally known as The National Institute for Automotive Service Excellence (NIASE), today's ASE was founded in 1972 as a nonprofit, independent entity dedicated to improving the quality of automotive service and repair through the voluntary testing and certification of automotive technicians. Until that time, consumers had no way of distinguishing between competent and incompetent automotive technicians. In the mid-1960s and early 1970s, efforts were made by several automotive industry affiliated associations to respond to this need. Though the associations were nonprofit, many regarded certification test fees merely as a means of raising additional operating capital. Also, some associations, having a vested interest, produced test scores heavily weighted in the favor of its members.

NIASE

From these efforts a new independent, nonprofit association, the National Institute for Automotive Service Excellence (NIASE), was established much to the credit of two educators, George R. Kinsler, Director of Program Development for the Wisconsin Board of Vocational and Adult Education in Madison, WI, and Myron H. Appel, Division Chairman at Cypress College in Cypress, CA.

Early efforts were to encourage voluntary certification in four general areas:

TEST AREA	TITLES
I. Engine	Engines, Engine Tune-Up, Block Assembly, Cooling and Lube Systems, Induction, Ignition, and Exhaust
II. Transmission	Manual Transmissions, Drive Line and Rear Axles, and Automatic Transmissions
III. Brakes and Suspension	Brakes, Steering, Suspension, and Wheels
IV. Electrical/Air Conditioning	Body/Chassis, Electrical Systems, Heating, and Air Conditioning

In early NIASE tests, Mechanic A, Mechanic B type questions were used. Over the years the trend has not changed, but in mid-1984 the term was changed to Technician A, Technician B to better emphasize sophistication of the skills needed to perform successfully in the modern motor vehicle industry. In certain tests the term used is Estimator A/B, Painter A/B, or Parts Specialist A/B. At about that same time, the logo was changed from "The Gear" to "The Blue Seal," and the organization adopted the acronym ASE for Automotive Service Excellence.

Since those early beginnings, several other related trades have been added. ASE now administers a comprehensive series of certification exams for automotive and light truck repair technicians, medium and heavy truck repair technicians, alternate fuels technicians, engine machinists, collision repair technicians, school bus repair technicians, and parts specialists.

The Series and Individual Tests

- Automotive and Light Truck Technician; consisting of: Engine Repair—Automatic Transmission/Transaxle—Manual Drive Train and Axles—Suspension and Steering—Brakes—Electrical/Electronic Systems—Heating and Air Conditioning—Engine Performance
- Medium and Heavy Truck Technician; consisting of: Gasoline Engines—Diesel Engines—Drive Train—Brakes—Suspension and Steering—Electrical/Electronic Systems—HVAC—Preventive Maintenance Inspection
- Alternate Fuels Technician; consisting of: Compressed Natural Gas Light Vehicles
- Advanced Series; consisting of: Automobile Advanced Engine Performance and Advanced Diesel Engine Electronic Diesel Engine Specialty
- Collision Repair Technician; consisting of: Painting and Refinishing—Non-Structural Analysis and Damage Repair—Structural Analysis and Damage Repair—Mechanical and Electrical Components—Damage Analysis and Estimating
- Engine Machinist Technician; consisting of: Cylinder Head Specialist—Cylinder Block Specialist—Assembly Specialist
- School Bus Repair Technician; consisting of: Body Systems and Special Equipment—Drive Train—Brakes—Suspension and Steering—Electrical/Electronic Systems—Heating and Air Conditioning
- Parts Specialist; consisting of: Automobile Parts Specialist—Medium/Heavy Truck Parts Specialist

A Brief Chronology

1970–1971	Original questions were prepared by a group of forty auto mechanics teachers from public secondary schools, technical institutes, community colleges, and private vocational schools. These questions were then professionally edited by testing specialists at Educational Testing Service (ETS) at Princeton, New Jersey, and thoroughly reviewed by training specialists associated with domestic and import automotive companies.
1971	July: About eight hundred mechanics tried out the original test questions at experimental test administrations.
1972	November and December: Initial NIASE tests administered at 163 test centers. The original automotive test series consisted of four tests containing eighty questions each. Three hours were allotted for each test. Those who passed all four tests were designated Certified General Auto Mechanic (GAM).
1973	April and May: Test 4 was increased to 120 questions. Time was extended to four hours for this test. There were now 182 test centers. Shoulder patch insignias were made available.

The History of ASE

	November: Automotive series expanded to five tests. Heavy-Duty Truck series of six tests introduced.
1974	November: Automatic Transmission (Light Repair) test modified. Name changed to Automatic Transmission.
1975	May: Collision Repair series of two tests is introduced.
1978	May: Automotive recertification testing is introduced.
1979	May: Heavy-Duty Truck recertification testing is introduced.
1980	May: Collision Repair recertification testing is introduced.
1982	May: Test administration providers switched from Educational Testing Service (ETS) to American College Testing (ACT). Name of Automobile Engine Tune-Up test changed to Engine Performance test.
1984	May: New logo was introduced. ASE's "The Blue Seal" replaced NIASE's "The Gear." All reference to Mechanic A, Mechanic B was changed to Technician A, Technician B.
1990	November: The first of the Engine Machinist test series was introduced.
1991	May: The second test of the Engine Machinist test series was introduced. November: The third and final Engine Machinist test was introduced.
1992	May: Name of Heavy-Duty Truck Test series changed to Medium/Heavy Truck test series.
1993	May: Automotive Parts Specialist test introduced. Collision Repair expanded to six tests. Light Vehicle Compressed Natural Gas test introduced. Limited testing begins in English-speaking provinces of Canada.
1994	May: Advanced Engine Performance Specialist test introduced.
1996	May: First three tests for School Bus Technician test series introduced. November: A Collision Repair test is added.
1997	May: A Medium/Heavy Truck test is added.
1998	May: A diesel advanced engine test is introduced: Electronic Diesel Engine Diagnosis Specialist. A test is added to the School Bus test series.

By the Numbers

Following are the approximate number of ASE technicians currently certified by category. The numbers may vary from time to time but are reasonably accurate for any given period. More accurate data may be obtained from ASE, which provides updates twice each year, in May and November after the Spring and Fall test series.

There are more than 338,000 Automotive Technicians with over 87,000 at Master Technician (MA) status. There are 47,000 Truck Technicians with over 19,000 at Master Technician (MT) status. There are 46,000 Collision Repair/Refinish Technicians with 7,300 at Master Technician (MB) status. There are 1,200 Estimators. There are 6,700 Engine Machinists with over 2,800 at Master Machinist Technician (MM) status. There are also 28,500 Automobile Advanced Engine Performance Technicians and over 2,700 School Bus Technicians for a combined total of more than 403,000 Repair Technicians. To this number, add over 22,000 Automobile Parts Specialists, and over 2,000 Truck Parts Specialists for a combined total of over 24,000 parts specialists.

There are over 6,400 ASE Technicians holding both Master Automotive Technician and Master Truck Technician status, of which 350 also hold Master Body Repair status. Almost 200 of these Master Technicians also hold Master Machinist status and five Technicians are certified in all ASE specialty areas.

Almost half of ASE certified technicians work in new vehicle dealerships (45.3 percent). The next greatest number work in independent garages with 19.8 percent. Next is tire dealerships with 9 percent, service stations at 6.3 percent, fleet shops at 5.7 percent, franchised volume retailers at 5.4 percent, paint and body shops at 4.3 percent, and specialty shops at 3.9 percent.

Of over 400,000 automotive technicians on ASE's certification rosters, almost 2,000 are female. The number of female technicians is increasing at a rate of about 20 percent each year. Women's increasing interest in the automotive industry is further evidenced by the fact that, according to the National Automobile Dealers Association (NADA), they influence 80 percent of the decisions of the purchase of a new automobile and represent 50 percent of all new car purchasers. Also, it is interesting to note that 65 percent of all repair and maintenance service customers are female.

The typical ASE certified technician is 36.5 years of age, is computer literate, deciphers a half-million pages of technical manuals, spends one hundred hours per year in training, holds four ASE certificates, and spends about $27,000 for tools and equipment. Twenty-seven percent of today's skilled ASE certified technicians attended college, many having earned an Associate of Science degree in Automotive Technology.

ASE

ASE's mission is to improve the quality of vehicle repair and service in the United States through the testing and certification of automotive repair technicians. Prospective candidates register for and take one or more of ASE's thirty-three exams. The tests are grouped into specialties for automobile, medium/heavy truck, school bus, and collision repair technicians as well as engine machinists, alternate fuels technicians, and parts specialists.

Upon passing at least one exam and providing proof of two years of related work experience, the technician becomes ASE certified. A technician who passes a series of exams earns ASE Master Technician status. An automobile technician, for example, must pass eight exams for this recognition.

The tests, conducted twice a year at over seven hundred locations around the country, are administered by American College Testing (ACT). They stress real-world diagnostic and repair problems. Though a good knowledge of theory is helpful to the technician in answering many of the questions, there are no questions specifically on theory. Certification is valid for five years. To retain certification, the technician must be retested to renew his or her certificate.

The automotive consumer benefits because ASE certification is a valuable yardstick by which to measure the knowledge and skills of individual technicians, as well as their commitment to their chosen profession. It is also a tribute to the repair facility employing ASE certified technicians. ASE certified technicians are permitted to wear blue and white ASE shoulder insignia, referred to as the "Blue Seal of Excellence," and carry credentials listing their areas of expertise. Often employers display their technicians' credentials in the customer waiting area. Customers look for facilities that display ASE's Blue Seal of Excellence logo on outdoor signs, in the customer waiting area, in the telephone book (Yellow Pages), and in newspaper advertisements.

The tests stress repair knowledge and skill. All test takers are issued a score report. In order to earn ASE certification, a technician must pass one or more of the exams and present proof of two years of relevant hands-on work experience. ASE certifications are valid for five years, after which time technicians must retest in order to keep up with changing technology and to remain in the ASE program. A nominal registration and test fee is charged.

To become part of the team that wears ASE's Blue Seal of Excellence®, please contact:

National Institute for Automotive Service Excellence
13505 Dulles Technology Drive
Herndon, VA 20171-3421

2 Take and Pass Every ASE Test

ASE Testing

Participating in an Automotive Service Excellence (ASE) voluntary certification program gives you a chance to show your customers that you have the "know-how" needed to work on today's modern vehicles. The ASE certification tests allow you to compare your skills and knowledge to the automotive service industry's standards for each specialty area.

If you are the "average" automotive technician taking this test, you are in your mid-thirties and have not attended school for about fifteen years. That means you probably have not taken a test in many years. Some of you, on the other hand, have attended college or taken postsecondary education courses and may be more familiar with taking tests and with test-taking strategies. There is, however, a difference in the ASE test you are preparing to take and the educational tests you may be accustomed to.

Who Writes the Questions?

The questions on an educational test are generally written, administered, and graded by an educator who may have little or no practical hands-on experience in the test area. The questions on all ASE tests are written by service industry experts familiar with all aspects of the subject area. ASE questions are entirely job-related and designed to test the skills that you need to know on the job.

The questions originate in an ASE "item-writing" workshop where service representatives from domestic and import automobile manufacturers, parts and equipment manufacturers, and vocational educators meet in a workshop setting to share their ideas and translate them into test questions. Each test question written by these experts is reviewed by all of the members of the group. The questions deal with the practical problems of diagnosis and repair that are experienced by technicians in their day-to-day hands-on work experiences.

All of the questions are pretested and quality-checked in a nonscoring section of tests by a national sample of certifying technicians. The questions that meet ASE's high standards of accuracy and quality are then included in the scoring sections of future tests. Those questions that do not pass ASE's stringent tests are sent back to the workshop or are discarded. ASE's tests are monitored by an independent proctor and are administered and machine-scored by an independent provider, American College Testing (ACT). All ASE tests have a three-year revision cycle.

Testing

If you think about it, we are actually tested on about everything we do. As infants, we were tested to see when we could turn over and crawl, later when we could walk or talk. As adolescents, we were tested to determine how well we learned the material presented in school and in how we demonstrated our accomplishments on the athletic field. As working adults, we are tested by our supervisors on how well we have completed an assignment or project. As nonworking adults, we are tested by our family on everyday activities, such as housekeeping or preparing a meal. Testing, then, is one of those facts of life that begins in the cradle and follows us to the grave.

Testing is an important fact of life that helps us to determine how well we have learned our trade. Also, tests often help us to determine what opportunities will be available to us in the future. To become ASE certified, we are required to take a test in every subject in which we wish to be recognized.

Be Test-Wise

In spite of the widespread use of tests, most technicians are not very test-wise. An ability to take tests and score well is a skill that must be acquired. Without this knowledge, the most intelligent and prepared technician may not do well on a test.

We will discuss some of the basic procedures necessary to follow in order to become a test-wise technician. Assume, if you will, that you have done the necessary study and preparation to score well on the ASE test.

Different approaches should be used for taking different types of tests. The different basic types of tests include: essay, objective, multiple-choice, fill-in-the-blank, true-false, problem solving, and open book. All ASE tests are of the four-part multiple-choice type.

Before discussing the multiple-choice type test questions, however, there are a few basic principles that should be followed before taking any test.

Before the Test

Do not arrive late. Always arrive well before your test is scheduled to begin. Allow ample time for the unexpected, such as traffic problems, so you will arrive on time and avoid the unnecessary anxiety of being late.

Always be certain to have plenty of supplies with you. For an ASE test, three or four sharpened soft lead (#2) pencils, a pocket pencil sharpener, erasers, and a watch are all that are required.

Do not listen to pretest chatter. When you arrive early, you may hear other technicians testing each other on various topics or making their best guess as to the probable test questions. At this time, it is too late to add to your knowledge. Also the rhetoric may only confuse you. If you find it bothersome, take a walk outside the test room to relax and loosen up.

Read and listen to all instructions. It is important to read and listen to the instructions. Make certain that you know what is expected of you. Listen carefully to verbal instructions and pay particular attention to any written instructions on the test paper. Do not dive into answering questions only to find out that you have answered the wrong question by not following instructions carefully. It is difficult to make a high score on a test if you answer the wrong questions.

These basic principles have been violated in most every test ever given. Try to remember them. They are essential for success.

Objective Tests

A test is called an objective test if the same standards and conditions apply to everyone taking the test and there is only one correct answer to each question. Objective tests primarily measure your ability to recall information. A well-designed objective test can also test your ability to understand, analyze, interpret, and apply your knowledge. Objective tests include true-false, multiple-choice, fill-in-the-blank, and matching questions.

Objective questions, not generally encountered in a classroom setting, are frequently used in standardized examinations. Objective tests are easy to grade and also reduce the amount of paperwork necessary to administer. The objective tests are used in entry-level programs or when very large numbers are being tested. ASE's tests consist exclusively of four-part multiple-choice objective questions in all of their tests.

Taking an Objective Test

The principles of taking an objective test are somewhat different from those used in other types of tests. You should first quickly look over the test to determine the number of questions, but do not try to read through all of the questions. In an ASE test, there are usually between forty and eighty questions, depending on the subject matter. Read through each question before marking your answer. Answer the questions in the order they appear on the test. Leave the questions blank that you are not sure of and move on to the next question. You can return to those unanswered questions after you have finished the others. They may be easier to answer at a later time after your mind has had additional time to consider them on a subconscious level. In addition, you might find information in other questions that will help you to answer some of them.

Do not be obsessed by the apparent pattern of responses. For example, do not be influenced by a pattern like **d, c, b, a, d, c, b, a** on an ASE test.

There is also a lot of folk wisdom about taking objective tests. For example, there are those who would advise you to avoid response options that use certain words such as *all, none, always, never, must,* and *only,* to name a few. This, they claim, is because nothing in life is exclusive. They would advise you to choose response options that use words that allow for some exception, such as *sometimes, frequently, rarely, often, usually, seldom,* and *normally.* They would also advise you to avoid the first and last option (A and D) because test writers, they feel, are more comfortable if they put the correct answer in the middle (B and C) of the choices. Another recommendation often offered is to select the option that is either shorter or longer than the other three choices because it is more likely to be correct. Some would advise you to never change an answer since your first intuition is usually correct.

Although there may be a grain of truth in this folk wisdom, ASE test writers try to avoid them and so should you. There are just as many **A** answers as there are **B** answers, just as many **D** answers as **C** answers. As a matter of fact, ASE tries to balance the answers at about 25 percent per choice **A**, **B**, **C**, and **D**. There is no intention to use "tricky" words, such as outlined above. Put no credence in the opposing words "sometimes" and "never," for example. When used in an ASE type question, one or both may be correct; one or both may be incorrect.

There are some special principles to observe on multiple-choice tests. These tests are sometimes challenging because there are often several choices that may seem possible, and it may be difficult to decide on the correct choice. The best strategy, in this case, is to first determine the correct answer before looking at the options. If you see the answer you decided on, you should still examine the options to make sure that none seem more correct than yours. If you do not know or are not sure of the answer, read each option very carefully and try to eliminate those options that you know to be wrong. That way, you can often arrive at the correct choice through a process of elimination.

If you have gone through all of the test and you still do not know the answer to some of the questions, then guess. Yes, guess. You then have at least a 25 percent chance of being correct. If you leave the question blank, you have no chance. In ASE tests, there is no penalty for being wrong. As the late President Franklin D. Roosevelt once advised a group of students, "It is common sense to take a method and try it. If it fails, admit it frankly and try another. But above all, try something."

During the Test

Mark your bubble sheet clearly and accurately. One of the biggest problems an adult faces in test-taking, it seems, is in placing an answer in the correct spot on a bubble sheet. Make certain that you mark your answer for, say, question 21, in the space on the bubble sheet designated for the answer for question 21. A correct response in the wrong bubble will probably be wrong. Remember, the answer sheet is machine scored and can only "read" what you have bubbled in. Also, do not bubble in two answers for the same question. For example, if you feel the answer to a particular question is **A** but think it may be **C**, do not bubble in both choices. Even if either **A** or **C** is correct, a double answer will score as an incorrect answer. It's better to take a chance with your best guess.

Review Your Answers

If you finish answering all of the questions on a test ahead of time, go back and review the answers of those questions that you were not sure of. You can often catch careless errors by using the remaining time to review your answers.

Don't Be Distracted

At practically every test, some technicians will invariably finish ahead of time and turn their papers in long before the final call. Do not let them distract or intimidate you. Either they knew too little and could not finish the test, or they were very self-confident and thought they knew it all. Perhaps they were trying to impress the proctor or other technicians about how much they know. Often you may hear them later talking about the information they knew all the while but forgot to respond on their answer sheet.

Use Your Time Wisely

It is not wise to use less than the total amount of time that you are allotted for a test. If there are any doubts, take the time for review. Any product can usually be made better with some additional effort. A test is no exception. It is not necessary to turn in your test paper until you are told to do so.

Don't Cheat

Some technicians may try to use a "crib sheet" during a test. Others may attempt to read answers from another technician's paper. If you do that, you are unquestionably assuming that someone else has a correct answer. You probably know as much, maybe more, than anyone else in the test room. Trust yourself. If you're still not convinced, think of the consequences of being caught. Cheating is foolish. If you are caught, you have failed the test.

Be Confident

The first and foremost principle in taking a test is that you need to know what you are doing, to be test-wise. It will now be presumed that you are a test-wise technician and are now ready for some of the more obscure aspects of test-taking.

An ASE-style test requires that you use the information and knowledge at your command to solve a problem. This generally requires a combination of information similar to the way you approach problems in the real world. Most problems, it seems, typically do not fall into neat textbook cases. New problems are often difficult to handle, whether they are encountered inside or outside the test room.

An ASE test also requires that you apply methods taught in class as well as those learned on the job to solve problems. These methods are akin to a well-equipped tool box in the hands of a skilled technician. You have to know what tools to use in a particular situation, and you must also know how to use them. In an ASE test, you will need to be able to demonstrate that you are familiar with and know how to use the tools.

You should begin a test with a completely open mind. At times, however, you may have to move out of your normal way of thinking and be creative to arrive at a correct answer. If you have diligently studied for at least one week prior to the test, you have bombarded your mind with a wide assortment of information. Your mind will be working with this information on a subconscious level, exploring the interrelationships among various facts, principles, and ideas. This prior preparation should put you in a creative mood for the test.

In order to reach your full potential, you should begin a test with the proper mental attitude and a high degree of self-confidence. You should think of a test as an opportunity to document how much you know about the various tasks in your chosen profession. If you have been diligently studying the subject matter, you will be able to take your test in serenity because your mind will be well organized. If you are confident, you are more likely to do well because you have the proper mental attitude. If, on the other hand, your confidence is low, you are bound to do poorly. It is a self-fulfilling prophecy.

Perhaps you have heard athletic coaches talk about the importance of confidence when competing in sports. Mental confidence helps an athlete to perform at the highest level and gain an advantage over competitors. Taking a test is much like an athletic event. You are competing against yourself, in a certain sense, because you will be trying to approach perfection in determining your answers. As in any competition, you should aim your sights high and be confident that you can reach the apex.

Anxiety and Fear

Many technicians experience anxiety and fear at the very thought of taking a test. Many worry, become nervous, and even become ill at test time because of the fear of failure. Many often worry about the criticism and ridicule that may come from their employer, relatives, and peers. Some worry about taking a test because they feel that the stakes are very high. Those who spent a great amount of time studying may feel they must get a high grade to justify their efforts. The thought of not doing well can result in unnecessary worry. They become so worried, in fact, that their reasoning and thinking ability is impaired, actually bringing about the problem they wanted to avoid.

The fear of failure should not be confused with the desire for success. It is natural to become "psyched-up" for a test in contemplation of what is to come. A little emotion can provide a healthy flow of adrenaline to peak your senses and hone your mental ability. This improves your performance on the test and is a very different reaction from fear.

Most technician's fears and insecurities experienced before a test are due to a lack of self-confidence. Those who have not scored well on previous tests or have no confidence in their preparation are those most likely to fail. Be confident that you will do well on your test and your fears should vanish. You will know that you have done everything possible to realize your potential.

Getting Rid of Fear

If you have previously experienced fear of taking a test, it may be difficult to change your attitude immediately. It may be easier to cope with fear if you have a better understanding of what the test is about. A test is merely an assessment of how much the technician knows about a particular task area. Tests, then, are much less threatening when thought of in this manner. This does not mean, however, that you should lower your self-esteem simply because you performed poorly on a test.

You can consider the test essentially as a learning device, providing you with valuable information to evaluate your performance and knowledge. Recognize that no one is perfect. All humans make mistakes. The idea, then, is to make mistakes before the test, learn from them, and avoid repeating them on the test. Fortunately, this is not as difficult as it seems. Practical questions in this study guide include the correct answers to consider if you have made mistakes on the practice test. You should learn where you went wrong so you will not repeat them in the ASE test. If you learn from your mistakes, the stage is set for future growth.

If you understood everything presented up until now, you have the knowledge to become a test-wise technician, but more is required. To be a test-wise technician, you not only have to practice these principles, you have to diligently study in your task area.

Effective Study

The fundamental and vital requirement to induce effective study is a genuine and intense desire to achieve. This is more basic than any rule or technique that will be given here. The key requirement, then, is a driving motivation to learn and to achieve.

If you wish to study effectively, first develop a desire to master your studies and sincerely believe that you will master them. Everything else is secondary to such a desire.

First, build up definite ambitions and ideals toward which your studies can lead. Picture the satisfaction of success. The attitude of the technician may be transformed from merely getting by to an earnest and energetic effort. The best direct stimulus to change may involve nothing more than the deliberate planning of your time. Plan time to study.

Another drive that creates positive study is an interest in the subject studied. As an automotive technician, you can develop an interest in studying particular subjects if you follow these four rules:

1. Acquire information from a variety of sources. The greater your interest in a subject, the easier it is to learn about it. Visit your local library and seek books on the subject you are studying. When you find something new or of interest, make inexpensive photocopies for future study.
2. Merge new information with your previous knowledge. Discover the relationship of new facts to old known facts. Modern developments in automotive technology take on new interest when they are seen in relation to present knowledge.
3. Make new information personal. Relate the new information to matters that are of concern to you. The information you are now reading, for example, has interest to you as you think about how it can help.
4. Use your new knowledge. Raise questions about the points made by the book. Try to anticipate what the next steps and conclusions will be. Discuss this new knowledge, particularly the difficult and questionable points, with your peers.

You will find that when you study with eager interest, you will discover it is no longer work. It is pleasure and you will be fascinated in what you study. Studying can be like reading a novel or seeing a movie that overcomes distractions and requires no effort or willpower. You will discover that the positive relationship between interest and effort works both ways. Even though you perhaps began your studies with little or no interest, simply staying with it helped you to develop an interest in your studies.

Obviously, certain subject matter studies are bound to be of little or no interest, particularly in the beginning. Parts of certain studies may continue to be uninteresting. An honest effort to master those subjects, however, nearly always brings about some level of interest. If you appreciate the necessity and reward of effective studying, you will rarely be disappointed. Here are a few important hints for gaining the determination that is essential to carrying good conclusions into actual practice.

Make Study Definite

Decide what is to be studied and when it is to be studied. If the unit is discouragingly long, break it into two or more parts. Determine exactly what is involved in the first part and learn that. Only then should you proceed to the next part. Stick to a schedule.

The Urge to Learn

Make clear to yourself the relation of your present knowledge to your study materials. Determine the relevance with regard to your long-range goals and ambitions.

Turn your attention away from real or imagined difficulties as well as other things that you would rather be doing. Some major distractions are thoughts of other duties and of disturbing problems. These distractions can usually be put aside, simply shunted off by listing them in a notebook. Most technicians have found that by writing interfering thoughts down, their minds are freed from annoying tensions.

Adopt the most reasonable solution you can find or seek objective help from someone else for personal problems. Personal problems and worry are often causes of ineffective study. Sometimes there are no satisfactory solutions. Some manage to avoid the problems or to meet them without great worry. For those who may wish to find better ways of meeting their personal problems, the following suggestions are offered:

1. Determine as objectively and as definitely as possible where the problem lies. What changes are needed to remove the problem, and which changes, if any, can be made? Sometimes it is wiser to alter your goals than external conditions. If there is no perfect solution, explore the others. Some solutions may be better than others.

2. Seek an understanding confidant who may be able to help analyze and meet your problems. Very often, talking over your problems with someone in whom you have confidence and trust will help you to arrive at a solution.

3. Do not betray yourself by trying to evade the problem or by pretending that it has been solved. If social problem distractions prevent you from studying or doing satisfactory work, it is better to admit this to yourself. You can then decide what can be done about it.

Once you are free of interferences and irritations, it is much easier to stay focused on your studies.

Concentrate

To study effectively, you must concentrate. Your ability to concentrate is governed, to a great extent, by your surroundings as well as your physical condition. When absorbed in study, you must be oblivious to everything else around you. As you learn to concentrate and study, you must also learn to overcome all distractions. There are three kinds of distractions you may face:

1. Distractions in the surrounding area, such as motion, noise, and the glare of lights. The sun shining through a window on your study area, for example, can be very distracting.

    Some technicians find that, for effective study, it is necessary to eliminate visual distractions as well as noises. Others find that they are able to tolerate moderate levels of auditory or visual distraction.

    Make sure your study area is properly lighted and ventilated. The lighting should be adequate but should not shine directly into your eyes or be visible out of the corner of your eye. Also, try to avoid a reflection of the lighting on the pages of your book.

    Whether heated or cooled, the environment should be at a comfortable level. For most, this means a temperature of 78°F–80°F (25.6°C–26.7°C) with a relative humidity of 45 to 50 percent.

2. Distractions arising from your body, such as a headache, fatigue, and hunger. Be in good physical condition. Eat wholesome meals at regular times. Try to eat with your family or friends whenever possible. Mealtime should be your recreational period. Do not eat a heavy meal for lunch, and do not resume studies immediately after eating lunch. Just after lunch, try to get some regular exercise, relaxation, and recreation. A little exercise on a regular basis is much more valuable than a lot of exercise only on occasion.
3. Distractions of irrelevant ideas, such as how to repair the garden gate, when you are studying for an automotive-related test.

The problems associated with study are no small matter. These problems of distractions are generally best dealt with by a process of elimination. A few important rules for eliminating distractions follow.

Get Sufficient Sleep

You must get plenty of rest even if it means dropping certain outside activities. Avoid cutting in on your sleep time; you will be rewarded in the long run. If you experience difficulty going to sleep, do something to take your mind off your work and try to relax before going to bed. Some suggestions that may help include a little reading, a warm bath, a short walk, a conversation with a friend, or writing that overdue letter to a distant relative. If sleeplessness is an ongoing problem, consult a physician. Do not try any of the sleep remedies on the market, particularly if you are on medication, without approval of your physician.

If you still have difficulty studying, a final rule may help. Sit down in a favorable place for studying, open your books, and take out your pencil and paper. In a word, go through the motions.

Arrange Your Area

Arrange your chair and work area. To avoid strain and fatigue, whenever possible, shift your position occasionally. Try to be comfortable; however, avoid being too comfortable. It is nearly impossible to study rigorously when settled back in a large easy chair or reclining leisurely on a sofa.

When studying, it is essential to have a plan of action, a time to work, a time to study, and a time for pleasure. If you schedule your day and adhere to the schedule, you will eliminate most of your efforts and worries. A plan that is followed, then, soon becomes the easy and natural routine of the day. Most technicians find it useful to have a definite place and time to study. A particular table and chair should always be used for study and intellectual work. This place will then come to mean study. To be seated in that particular location at a regularly scheduled time will automatically lead you to assume a readiness for study.

Don't Daydream

Daydreaming or mind-wandering is an enemy of effective study. Daydreaming is frequently due to an inadequate understanding of words. Use the Glossary or a dictionary to look up the troublesome word. Another frequent cause of daydreaming is a deficient background in the present subject matter. When this is the problem, go back and review the subject matter to obtain the necessary foundation. Just one hour of concentrated study is equivalent to ten hours with frequent lapses of daydreaming. Be on guard against mind-wandering, and pull yourself back into focus on every occasion.

Study Regularly

A system of regularity in study is believed by many scholars to be the secret of success. The daily time schedule must, however, be determined on an individual basis. You must decide how many hours of each day you can devote to your studies. Few technicians really are aware of where their leisure time is spent. An accurate account of how your days are presently being spent is an important first step toward creating an effective daily schedule.

	Weekly Schedule						
	Sun	Mon	Tues	Wed	Thu	Fri	Sat
6:00							
6:30							
7:00							
7:30							
8:00							
8:30							
9:00							
9:30							
10:00							
10:30							
11:00							
11:30							
NOON							
12:30							
1:00							
1:30							
2:00							
2:30							
3:00							
3:30							
4:00							
4:30							
5:00							
5:30							
6:00							
6:30							
7:00							
7:30							
8:00							
8:30							
9:00							
9:30							
10:00							
10:30							
11:00							
11:30							

The convenient form is for keeping an hourly record of your week's activities. If you fill in the schedule each evening before bedtime, you will soon gain some interesting and useful facts about yourself and your use of your time. If you think over the causes of wasted time, you can determine how you might better spend your time. A practical schedule can be set up by using the following steps.

1. Mark your fixed commitments, such as work, on your schedule. Be sure to include classes and clubs. Do you have sufficient time left? You can arrive at an estimate of the time you need for studying by counting the hours used during the present week. An often-used formula, if you are taking classes, is to multiply the number of hours you spend in class by two. This provides time for class studies. This is then added to your work hours. Do not forget time allocation for travel.

2. Fill in your schedule for meals and studying. Use as much time as you have available during the normal workday hours. Do not plan, for example, to do all of your studying between 11:00 P.M. and 1:00 A.M. Try to select a time for study that you can use every day without interruption. You may have to use two or perhaps three different study periods during the day.

3. List the things you need to do within a time period. A one-week time frame seems to work well for most technicians. The question you may ask yourself is: "What do I need to do to be able to walk into the test next week, or next month, prepared to pass?"

4. Break down each task into smaller tasks. The amount of time given to each area must also be settled. In what order will you tackle your schedule? It is best to plan the approximate time for your assignments and the order in which you will do them. In this way, you can avoid the difficulties of not knowing what to do first and of worrying about the other things you should be doing.

5. List your tasks in the empty spaces on your schedule. Keep some free time unscheduled so you can deal with any unexpected events, such as a dental appointment. You will then have a tentative schedule for the following week. It should be flexible enough to allow some units to be rearranged if necessary. Your schedule should allow time off from your studies. Some use the promise of a planned recreational period as a reward for motivating faithfulness to a schedule. You will more likely lose control of your schedule if it is packed too tightly.

Keep a Record

Keep a record of what you actually do. Use the knowledge you gain by keeping a record of what you are actually doing so you can create or modify a schedule for the following week. Be sure to give yourself credit for movement toward your goals and objectives. If you find that you cannot study productively at a particular hour, modify your schedule so as to correct that problem.

Scoring the ASE Test

You can gain a better perspective about tests if you know and understand how they are scored. ASE's tests are scored by American College Testing (ACT), a nonpartial, nonbiased organization having no vested interest in ASE or in the automotive industry. Each question carries the same weight as any other question. For example, if there are fifty questions, each is worth 2 percent of the total score. The passing grade is 70 percent. That means you must correctly answer thirty-five of the fifty questions to pass the test.

Understand the Test Results

The test results can tell you:
- where your knowledge equals or exceeds that needed for competent performance, or
- where you might need more preparation.

The test results *cannot* tell you:
- how you compare with other technicians, or
- how many questions you answered correctly.

Your ASE test score report will show the number of correct answers you got in each of the content areas. These numbers provide information about your performance in each area of the test. However, because there may be a different number of questions in each area of the test, a high percentage of correct answers in an area with few questions may not offset a low percentage in an area with many questions.

It may be noted that one does not "fail" an ASE test. The technician that does not pass is simply told "More Preparation Needed." Though large differences in percentages may indicate problem areas, it is important to consider how many questions were asked in each area. Since each test evaluates all phases of the work involved in a service specialty, you should be prepared in each area. A low score in one area could keep you from passing an entire test.

Note that a typical test will contain the number of questions indicated above each content area's description. For example:

Gasoline Engines (Test T1)

Content Area	Questions	Percent of Test
A. General Engine Diagnosis	15	19%
B. Cylinder Head and Valve Train Diagnosis and Repair	8	10%
C. Engine Block Diagnosis and Repair	8	10%
D. Lubrication and Cooling Systems Diagnosis and Repair	8	10%
E. Ignition System Diagnosis and Repair	11	14%
F. Fuel and Exhaust Systems Diagnosis and Repair	10	12%
G. Battery and Starting Systems Diagnosis and Repair	7	9%
H. Emissions Control Systems Diagnosis and Repair	7	9%
I. Computerized Engine Controls Diagnosis and Repair	6	7%
Total	*80	100%

***Note:** *The test could contain up to ten additional questions that are included for statistical research purposes only. Your answers to these questions will not affect your score, but since you do not know which ones they are, you should answer all questions in the test. The five-year Recertification Test will cover the same content areas as those listed above. However, the number of questions in each content area of the Recertification Test will be reduced by about one-half.*

"Average"

There is no such thing as average. You cannot determine your overall test score by adding the percentages given for each task area and dividing by the number of areas. It doesn't work that way because there generally are not the same number of questions in each task area. A task area with twenty questions, for example, counts more toward your total score than a task area with ten questions.

So, How Did You Do?

Your test report should give you a good picture of your results and a better understanding of your task areas of strength and weakness.

If you fail to pass the test, you may take it again at any time it is scheduled to be administered. You are the only one who will receive your test score. Test scores will not be given over the telephone by ASE nor will they be released to anyone without your written permission.

3 Are You Sure You're Ready for Test T1?

Pretest

The purpose of this pretest is to determine the amount of review that you may require prior to taking the ASE medium/heavy truck test: Gasoline Engines (Test T1). If you answer all of the pretest questions correctly, complete the sample test in section 5 along with the additional test questions in section 6.

If two or more of your answers to the pretest questions are wrong, study section 4: An Overview of the System before continuing with the sample test and additional test questions.

The pretest answers and explanations are located at the end of the pretest.

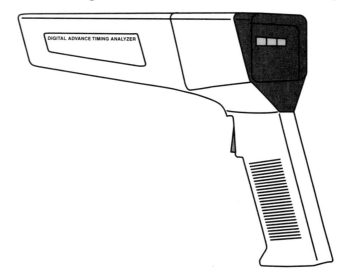

1. The timing light like the one in the figure above has the capability to do which of the following:
   A. to be used as a trouble light.
   B. to check spark advance.
   C. to be used as an oscilloscope on the secondary ignition circuit.
   D. to check the primary circuit ignition timing.

2. Valve seats are typically ground to an angle of:
   A. 15 or 20.
   B. 20 or 30.
   C. 30 or 45.
   D. 45 or 60.

3. When installing RTV sealer:
   A. the components to be sealed should be washed with an oil-base solvent.
   B. the RTV bead 1/8 inch wide should be placed in the center of the sealing surface.
   C. the RTV bead should be placed on one side of any bolt holes.
   D. the RTV bead should be allowed to dry for 10 minutes before component installation.

4. The thermostat is stuck open on a port fuel-injected engine. The most likely result of this problem is:
   A. a rich air-fuel ratio.
   B. a lean air-fuel ratio.
   C. excessive fuel pressure.
   D. engine overheating.

5. A port fuel-injected engine has a steady "puff" noise in the exhaust with the engine idling. The most likely cause of this problem is:
   A. a burned exhaust valve.
   B. excessive fuel pressure.
   C. a restricted fuel return line.
   D. a sticking fuel pump check valve.

6. Which of the following would LEAST likely cause an engine misfire?
   A. Intake manifold leak
   B. Exhaust manifold leak
   C. Defective spark plugs
   D. Defective coil

7. Technician A says the vibration damper counterbalances the back-and-forth twisting motion of the crankshaft each time a cylinder fires. Technician B says if the seal contact area on the vibration damper hub is scored, the damper assembly must be replaced. Who is right?
   A. A only
   B. B only
   C. Both A and B
   D. Neither A nor B

8. Which of the following is LEAST likely a normal oil pump component measurement?
   A. Outer rotor to housing clearance
   B. Clearance between the rotors
   C. Inner and outer rotor thickness
   D. Inner rotor diameter

9. An excessively high coolant level in the recovery reservoir may be caused by any of these problems **EXCEPT:**
   A. restricted radiator tubes.
   B. a thermostat that is stuck open.
   C. a loose water pump impeller.
   D. an inoperative electric-drive cooling fan.

10. All of these are methods of measuring engine V-belt tension **EXCEPT:**
    A. a belt tension gauge.
    B. measure the amount of belt deflection.
    C. visually see if the belt is contacting the bottom of the pulley.
    D. measure the length of the belt compared to a new one.

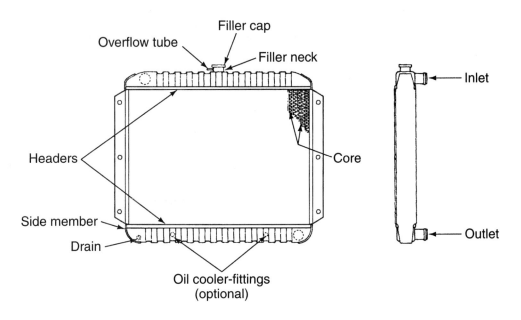

11. In the figure above, what type of radiator is shown?
    A. Downdraft
    B. Updraft
    C. Crossflow
    D. Downflow

12. While discussing cylinder measurement, Technician A says the cylinder taper is the difference between the cylinder diameter at the top of the ring travel compared to the cylinder diameter at the center of the ring travel. Technician B says cylinder out-of-round is the difference between the axial cylinder bore diameter at the top of the ring travel compared to the thrust cylinder bore diameter at the bottom of the ring travel. Who is right?
    A. A only
    B. B only
    C. Both A and B
    D. Neither A nor B

Answers to the Test Questions for the Pretest

1. B, 2. C, 3. B, 4. A, 5. A, 6. B, 7. A, 8. D, 9. B, 10. D, 11. D, 12. D

Explanations to the Answers for the Pretest

Question #1
Answer A is wrong. Timing lights cannot be used as trouble lights.
Answer B is correct. This timing light can check spark advance.
Answer C is wrong. Timing lights cannot be used as oscilloscopes.
Answer D is wrong. All timing checks occur on the secondary side.

Question #2
Answer A is wrong. 15 or 20 degrees is too wide.
Answer B is wrong. 20 to 30 degrees is too wide.
Answer C is correct. Seats are typically ground to either a 30 or 45 degree angle.
Answer D is wrong. 45 to 60 is too narrow of an angle.

Question #3
Answer A is wrong. Chlorinated solvent must be used to clean RTV sealed components.
Answer B is correct. The RTV bead 1/8 inch wide should be placed in the center of the sealing surface.
Answer C is wrong. The RTV bead must surround any bolt holes.
Answer D is wrong. RTV sealer cures in five minutes so components using this sealer must be assembled quickly.

Question #4
Answer A is correct. A stuck open thermostat will lower the temperature of the engine, which decreases operating temperature and can cause a rich air-fuel ratio.
Answer B is wrong. Higher temperatures can cause a lean air-fuel ratio.
Answer C is wrong. A stuck open thermostat has no effect on fuel pressure.
Answer D is wrong. A stuck open thermostat will not cause overheating.

Question #5
Answer A is correct. A burned exhaust valve causes a "puff" noise in the exhaust and is the most likely cause.
Answer B is wrong. High fuel pressure causes a rich air-fuel ratio with no puff.
Answer C is wrong. A restricted return fuel line causes a rich air-fuel ratio without a puff.
Answer D is wrong. A sticking fuel pump check valve may cause hard starting, but will not give a puff out the exhaust.

Question #6
Answer A is wrong. Intake manifold leaks commonly cause misfires.
Answer B is correct. An exhaust manifold leak is the least likely cause of a misfire.
Answer C is wrong. Spark plugs are often the cause of misfires.
Answer D is wrong. A defective coil can cause misfires.

Question #7
Answer A is correct. The vibration damper counterbalances the back-and-forth twisting crankshaft motion.
Answer B is wrong. If the damper seal contact area is scored, the damper hub may be machined and a sleeve installed to provide a new seal contact area.
Answer C is wrong.
Answer D is wrong.

Question #8
Answer A is wrong. Outer rotor to housing clearance is measured.
Answer B is wrong. Clearance between the rotors is measured.
Answer C is wrong. Inner and outer rotor thickness is measured.
Answer D is correct. This is the least likely because inner rotor diameter is normally not measured.

Question #9
Answer A is wrong. Restricted radiator tubes will decrease the cooling effectiveness of the radiator causing overexpansion of the coolant.
Answer B is correct. A stuck open thermostat will prevent the engine coolant from ever warming up.
Answer C is wrong. If the water pump impeller is loose, it will inhibit coolant flow prohibiting the cooling effects of the radiator causing the engine to overheat.
Answer D is wrong. An inoperative electric-drive cooling fan will inhibit airflow past the radiator prohibiting the cooling effects of the radiator causing the engine to overheat.

Question #10
Answer A is wrong. A belt tension gauge can be used.
Answer B is wrong. The amount of belt deflection can be measured.
Answer C is wrong. It can be visually seen if the belt is contacting the bottom of the pulley.
Answer D is correct. You do not measure the length of the belt compared to a new one.

Question #11
Answer A is wrong. There is no such radiator as downdraft.
Answer B is wrong. There is no such radiator as updraft.
Answer C is wrong. This is not the type of radiator shown.
Answer D is correct. A downflow radiator is shown.

Question #12
Answer A is wrong. Cylinder taper is the difference between the cylinder diameter at the top of the ring travel compared to the diameter at the bottom of the ring travel.
Answer B is wrong. Cylinder out-of-round is the difference between the axial bore diameter and the thrust bore diameter at the same cylinder position.
Answer C is wrong.
Answer D is correct.

Types of Questions

ASE certification tests are often thought of as being tricky. They may seem to be tricky if you do not completely understand what is being asked. The following examples will help you recognize certain types of ASE questions and avoid common errors.

Each test is made up of forty to eighty multiple-choice questions. Multiple-choice questions are an efficient way to test knowledge. To answer them correctly, you must think about each choice as a possibility, and then choose the one that best answers the question. To do this, read each word of the question carefully. Do not assume you know what the question is about until you have finished reading it.

Multiple-Choice Questions

One type of multiple-choice question has three wrong answers and one correct answer. The wrong answers, however, may be almost correct, so be careful not to jump at the first answer that seems to be correct. If all the answers seem to be correct, choose the answer that is the most correct. If you readily know the answer, this kind of question does not present a problem. If you are unsure of the answer, analyze the question and the answers. For example:

Question 1:

A heavy thumping machine gun sounding type noise occurs with the engine idling, but the oil pressure is normal. The noise may be caused by:
A. worn pistons.
B. loose flywheel bolts.
C. worn main bearings.
D. loose camshaft bearings.

Analysis:

Answer A is wrong. Piston noise would diminish after the engine was at operating temperature.
Answer B is correct. Loose flywheel bolts can cause this thumping noise.
Answer C is wrong. If the main bearings were in question, the oil pressure would indicate there was a problem.
Answer D is wrong. If the main bearings were in question, the oil pressure would indicate there was a problem.

EXCEPT Questions

Another type of question used on ASE tests has answers that are all correct except one. The correct answer for this type of question is the answer that is wrong. The word "EXCEPT" will always be in capital letters. You must identify which of the choices is the wrong answer. If you read quickly through the question, you may overlook what the question is asking and answer the question with the first correct statement. This will make your answer wrong. An example of this type of question and the analysis is as follows:

Question 2:
> The following are normal oil pump component measurements **EXCEPT:**
> A. inner rotor diameter.
> B. clearance between the rotors.
> C. inner and outer rotor thickness.
> D. outer rotor to housing clearance.

Analysis:

Answer A is correct. Inner rotor diameter is not something you normally measure.
Answer B is wrong. Clearance is measured between rotors.
Answer C is wrong. Inner and outer rotor thickness is measured.
Answer D is wrong. Outer rotor to housing clearance is measured.

Technician A, Technician B Questions

The type of question that is most popularly associated with an ASE test is the "Technician A says... Technician B says... Who is right?" type. In this type of question, you must identify the correct statement or statements. To answer this type of question correctly, you must carefully read each technician's statement and judge it on its own merit to determine if the statement is true.

Typically, this type of question begins with a statement about some analysis or repair procedure. This is followed by two statements about the cause of the problem, proper inspection, identification, or repair choices. You are asked whether the first statement, the second statement, both statements, or neither statement is correct. Analyzing this type of question is a little easier than the other types because there are only two ideas to consider although there are still four choices for an answer.

Technician A... Technician B questions are really double-true-false questions. The best way to analyze this kind of question is to consider each technician's statement separately. Ask yourself, is A true or false? Is B true or false? Then select your answer from the four choices. An important point to remember is that an ASE Technician A... Technician B question will never have Technician A and B directly disagreeing with each other. That is why you must evaluate each statement independently. An example of this type of question and the analysis of it follows.

Question 3:
> A cooling system is pressurized with a pressure tester to locate a coolant leak. After 15 minutes, the tester gauge has dropped from 15 to 5 psi, and there are no visible signs of coolant leaks in the engine compartment. Technician A says the engine may have a leaking head gasket. Technician B says that the heater core may be leaking. Who is right?
> A. A only
> B. B only
> C. Both A and B
> D. Neither A nor B

Analysis:

Answer A is wrong. With no visible external signs, the engine may have a leaking head gasket.
Answer B is wrong. The heater core may be leaking.
Answer C is correct. Both technicians are right.
Answer D is wrong.

Questions with a Figure

About 10 percent of ASE questions will have a figure, as shown in the following example:

Question 4:

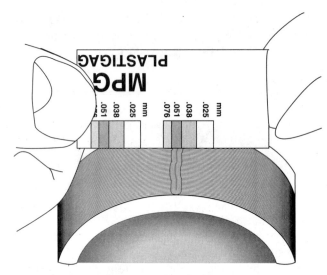

In the figure above, what is being performed?
A. Removing a scratch in the bearing with crocus cloth
B. Using a special tool to remove the bearing insert
C. Measuring the thickness of the crushed Plastigage®
D. Determining if the crankshaft journal has been machined

Analysis:
Answer A is wrong. If a scratch is deep enough to catch with a fingernail, the crankshaft needs machining.
Answer B is wrong. There is no special tool shown in the figure.
Answer C is correct. The technician is measuring the thickness of the crushed Plastigage® in the figure.
Answer D is wrong. To determine if the crankshaft has been machined, the bearing insert will indicate that it is oversized.

Most-Likely Questions

Most-likely questions are somewhat difficult because only one choice is correct while the other three choices are nearly correct. An example of a most-likely-cause question is as follows:

Question 5:

Which of the following exhaust conditions is most likely indication of an engine misfire?
A. Blue colored smoke from the tailpipe
B. Excessive rattling of the exhaust system
C. Cold exhaust temperature
D. Puffing or wheezing

Analysis:
Answer A is wrong. Blue colored smoke from the tailpipe usually indicates an internal oil leak.
Answer B is wrong. Excessive rattling of the exhaust system would only indicate a very severe misfire.
Answer C is wrong. Misfires cannot be determined by exhaust temperature.
Answer D is correct. Puffing or wheezing indicates an engine misfire.

LEAST-Likely Questions

Notice that in most-likely questions there is no capitalization. This is not so with least-likely type questions. For this type of question, look for the choice that would be the least likely cause of the described situation. Read the entire question carefully before choosing your answer. An example is as follows:

Question 6:
  Which of the following would LEAST likely cause an engine misfire?
  A. Intake manifold leak
  B. Exhaust manifold leak
  C. Defective spark plugs
  D. Defective coil
Analysis:
Answer A is wrong. Intake manifold leaks commonly cause misfires.
Answer B is correct. An exhaust manifold leak is the least likely cause of a misfire.
Answer C is wrong. Spark plugs are often the cause of misfires.
Answer D is wrong. A defective coil can cause misfires.

Summary

There are no four-part multiple-choice ASE questions having "none of the above" or "all of the above" choices. ASE does not use other types of questions, such as fill-in-the-blank, completion, true-false, word-matching, or essay. ASE does not require you to draw diagrams or sketches. If a formula or chart is required to answer a question, it is provided for you. There are no ASE questions that require you to use a pocket calculator.

Testing Time Length

An ASE test session is four hours and fifteen minutes. You may attempt from one to a maximum of four tests in one session. It is recommended, however, that no more than a total of 225 questions be attempted at any test session. This will allow for just over one minute for each question.

Visitors are not permitted at any time. If you wish to leave the test room, for any reason, you must first ask permission. If you finish your test early and wish to leave, you are permitted to do so only during specified dismissal periods.

Monitor Your Progress

You should monitor your progress and set an arbitrary limit to how much time you will need for each question. This should be based on the number of questions you are attempting. It is suggested that you wear a watch because some facilities may not have a clock visible to all areas of the room.

Registration

Test centers are assigned on a first-come, first-served basis. To register for an ASE certification test, you should enroll at least six weeks before the scheduled test date. This should provide sufficient time to assure you a spot in the test center. It should also give you enough time for study in preparation for the test. Test sessions are offered by ASE twice each year, in May and November, at over six hundred sites across the United States. Some tests that relate to emission testing also are given in August in several states.

To register, contact Automotive Service Excellence/American College Testing at:

ASE/ACT
P.O. Box 4007
Iowa City, IA 52243

4 An Overview of the System

Gasoline Engines (Test T1)

The following section includes the task areas and task lists for this test and a written overview of the topics covered in the test.

The task list describes the actual work you should be able to do as a technician that you will be tested on by the ASE. This is your key to the test and you should review this section carefully. We have based our sample test and additional questions upon these tasks, and the overview section will also support your understanding of the task list. ASE advises that the questions on the test may not equal the number of tasks listed; the task lists tell you what ASE expects you to know how to do and be ready to be tested on.

At the end of each question in the Sample Test and Additional Test Questions sections, a letter and number will be used as a reference back to this section for additional study. Note the following example: **Task C3.**

Task List

C. Engine Block Diagnosis and Repair (8 Questions)

Task C3 Inspect and measure cylinder walls for wear and damage; determine needed service.

Example:

1. While discussing cylinder measurement, Technician A says the cylinder taper is the difference between the cylinder diameter at the top of the ring travel compared to the cylinder diameter at the center of the ring travel. Technician B says cylinder out-of-round is the difference between the axial cylinder bore diameter at the top of the ring travel compared to the thrust cylinder bore diameter at the bottom of the ring travel. Who is right?
    A. A only
    B. B only
    C. Both A and B
    D. Neither A nor B

Question #1
Answer A is wrong because cylinder taper is the difference between the cylinder diameter at the top of the ring travel compared to the diameter at the bottom of ring travel.
Answer B is wrong because cylinder out-of-round is the difference between the axial bore diameter and the thrust bore diameter at the same cylinder position.
Answer C is wrong because both technicians are wrong.
Answer D is correct because neither technician is right.

Task List and Overview

A. General Engine Diagnosis (15 Questions)

Task A1 Listen to driver's complaint and road test vehicle; determine needed repairs.

The technician must be familiar with a basic diagnostic procedure such as the following: Listen carefully to the customer's complaint, and question the customer to obtain more information regarding the problem. Identify the complaint, and road test the vehicle, if necessary. Think of the possible causes of the problem. Perform diagnostic tests to locate the exact cause of the problem. Always start with the easiest, quickest test. After the repair has been made, be sure the customer's complaint is eliminated; road test the vehicle again if necessary.

Task A2 Inspect engine assembly for fuel, oil, coolant, and other leaks; determine needed repairs.

A technician must understand the basic fuel, lubricating, and cooling systems and components. The location of all possible leaks in these systems must be identified. Coolant leaks may be internal or external in relation to the engine. If a vacuum leak is sufficient to cause the engine to stall while idling, it is possible to locate the leak by listening for the whistle generated as airflow past the hole. Another method of vacuum leak detection is to spray carburetor cleaner around suspected areas while observing engine rpm with a tachometer. Since the carburetor cleaner is combustible, it will aid in enriching the mixture, which will increase engine speed.

Many times when trying to locate an engine oil leak, the engine has been leaking for an indefinite period and the engine oil has accumulated dirt. In this case, the engine must be steam cleaned to make a leak more visible. Another way to find an oil leak is to add a dye to the engine oil that reacts with an ultraviolet lamp.

Task A3 Check the level and the condition of fuel, oil, and coolant; determine needed repairs.

When checking the condition of engine oil, one observes the level on the dipstick. The color or the darkness of the oil is a key to determining whether you should change the oil. The darker in color, the greater amount of carbon particulate in the oil.

Always use caution when working with pressurized cooling systems. Engine coolant should be tested for freezing point and for acidity. You use a cooling system hydrometer with graduated temperature scale to check the freezing point of the coolant. Acidity can be checked using pH strips that are dipped into the coolant.

Task A4 Listen to engine noises; determine needed repairs.

Worn pistons and cylinders would cause a rapping noise while accelerating, and worn main bearings cause a thumping noise when you accelerate the engine. Loose camshaft bearings usually do not cause a noise unless severely worn.

The most common noise complaint is valve train noise. The technician identifies valve tappet noise by a light, regular clicking sound at twice the engine speed that comes from the upper portion of the engine. Dirty hydraulic lifters, lack of lubrication, and misadjusted valve clearance are some of the causes of valve train noise.

Ignition detonation and preignition can cause noises that can be mistaken for internal engine components. The sound is the result of a second flame that starts after the spark plug ignites. When the two flame fronts collide, one hears a loud explosion or knock. These conditions can be caused by internal engine components or by low quality fuel or a faulty ignition system. Other components that can create noise include flywheels, harmonic balancers, any belt driven component, torque converters, and motor mounts.

Task A5 Check color and quantify of the engine exhaust smoke; determine needed repairs.

Blue colored exhaust smoke indicates that an excessive amount of oil is entering the combustion chamber. Blue smoke during acceleration indicates worn piston rings. Blue smoke after startup indicates worn valve guides or seals. Black smoke indicates excessive fuel consumption; this is most likely a fuel system component concern. Gray or white smoke is an indication that coolant is entering the combustion chamber. The sound of the exhaust should be smooth and even. Puffing or wheezing exhaust or excessive rattling of the exhaust system itself indicates a misfire.

Task A6 Perform engine vacuum test; determine needed repairs.

The technician connects a vacuum gauge directly to the intake manifold to diagnose engine and related system conditions. When a vacuum gauge is connected to the intake manifold, the reading on the gauge should provide a steady reading between 17 and 22 inches of mercury (inch Hg) with the engine idling. Abnormal vacuum gauge readings indicate these problems:

- A low, steady reading indicates late ignition timing.
- If the vacuum gauge reading is steady and much lower than normal, the intake manifold has a significant leak.
- When the vacuum gauge pointer fluctuates between approximately 11 and 16 in. Hg on a carbureted engine at idle speed, the carburetor idle mixture screws require adjusting. On a fuel-injected engine, the injectors require cleaning or replacing.
- Burned or leaking valves cause a vacuum gauge fluctuation between 12 and 18 in. Hg.
- Weak valve springs result in a vacuum gauge fluctuation between 10 and 25 in. Hg.
- A leaking head gasket may cause a vacuum gauge fluctuation between 7 and 20 in. Hg.
- If the valves are sticking, the vacuum gauge fluctuates between 14 and 18 in. Hg.

If the vacuum gauge pointer drops to a very low reading when you accelerate the engine and hold a steady higher rpm, the catalytic converter or other exhaust system components are restricted.

Task A7 Perform cylinder power balance test; determine needed repairs.

If the cylinder is working normally, a noticeable rpm decrease occurs when the cylinder misfires. If there is very little rpm decrease when the analyzer causes a cylinder to misfire, the cylinder is not contributing to engine power. Under this condition the engine compression, ignition system, and fuel system should be checked to locate the cause of the problem. An intake manifold vacuum leak may cause a cylinder misfire with the engine idling or operating at low speed. If this problem exists, the misfire will disappear at a higher speed when the manifold vacuum decreases. When all the cylinders provide the specified rpm drop the cylinders are all contributing equally to the engine power?

Task A8 Perform cylinder compression tests; determine needed repairs.

You disable the ignition and fuel-injection system before proceeding with the compression test. During the compression test the engine is cranked through four compression strokes on each cylinder and the compression readings recorded. One interprets lower than specified compression readings as follows:

- Low compression readings on one or more cylinders indicate worn rings or valves, a blown head gasket, or a cracked cylinder head.
- A gradual buildup on the four compression readings on each stroke indicates worn rings, whereas little buildup on the four strokes usually is the result of a burned exhaust valve.

- When the compression readings on all the cylinders are even, but lower than the specified compression, you can suspect worn rings and cylinders.
- A leaking head gasket or cracked cylinder head causes low compression on two adjacent cylinders.
- Higher than specified compression usually indicates carbon deposits in the combustion chamber.
- A hole in the piston or an evenly burned exhaust valve will cause zero compression in a cylinder. If the zero compression reading is caused by a hole in the piston, the engine will have excessive blowby.

When the engine spins freely and compression in all cylinders is low, check the valve timing. If a cylinder compression reading is below specifications, you perform a wet test to determine if the valves, or rings, is the cause of the problem. You squirt approximately 2 or 3 teaspoons of engine oil through the spark plug opening into the cylinder, with the low compression reading. Crank the engine to distribute the oil around the cylinder wall and then retest the compression. If the compression reading improves considerably, the rings (or cylinders) are worn. When there is little change in the compression reading, one of the valves is leaking.

Task A9 Perform cylinder leakage tests; determine needed repairs.

During the leakage test a regulated amount of air from the shop air supply is forced into the cylinder with both exhaust and intake valves closed. The gauge on the leakage tester indicates the percentage of air escaping the cylinder. A gauge reading of 0 percent indicates there is no cylinder leakage; if the reading is 100 percent, the cylinder is not holding any air.

An excessive reading is one that exceeds 20 percent. Check for air escaping from the tailpipe, positive crankcase valve (PCV) opening, or the top of the throttle body or carburetor. Air escaping from the tailpipe indicates an exhaust valve leak. When the air is escaping out of the PCV valve opening, that is an indication that the piston rings are leaking. An intake valve is leaking if air is escaping from the top of the throttle body or carburetor. Remove the radiator cap and check the coolant for bubbles, which indicate a leaking head gasket or cracked cylinder head.

Task A10 Diagnose engine mechanical, ignition, or fuel problems with an engine oscilloscope and or analyzer (scan tool); determine needed repairs.

On many engine analyzers, ignition performance tests include primary circuit tests, secondary kilovolt (kV) tests, acceleration tests, scope patterns, and cylinder miss recall. Primary circuit tests include coil-input voltage, coil primary resistance, dwell, curb idle speed, and idle vacuum. The kV test measures the voltage required to start firing a spark plug. A high resistance problem in a spark plug or a spark plug wire causes higher firing kV, whereas a fouled spark plug or a cylinder with low compression results in a lower firing kV.

Some secondary kV tests include a snap kV test in which the analyzer directs the technician to accelerate the engine suddenly. When this action is taken, the firing kV should increase evenly on each cylinder. Some engine analyzers also display circuit gap for each cylinder. The circuit gap is the voltage to fire all the gaps in the secondary circuit, such as the rotor gap, but the spark plug gap is excluded. Some analyzers display the burn time for each cylinder with the secondary kV test. The burn time is the length of the spark line in milliseconds (ms). The average burn time should be 1 to 1.5 milliseconds.

B. Cylinder Head and Valve Train Diagnosis and Repair (8 Questions)

Task B1 **Inspect cylinder heads for cracks and gasket surface areas for warpage; check passage condition.**

When inspecting the cylinder head, it is normal if carbon buildup is a light even layer across the entire combustion chamber. If the carbon is excessive, the cylinder head should be thoroughly cleaned. This could be caused by worn valve guides, valve seals, or rings.

Once the initial cleaning has been completed, inspect the cylinder head for cracks and other obvious damage. Remember, not all cracks are visible to the eye. Therefore, it may be necessary to perform additional tests if a crack is suspected. Common locations for cracks are between the valve seats or around the spark plug hole. If the cylinder head has excessive damage, it may be more cost efficient to replace it instead of repairing it.

Check the cylinder head mating surface for texture and warpage. Use a straightedge with a feeler gauge to measure deck warpage. Warpage can occur in any direction on the head surface. Measure for warpage in three areas along the edges and three areas across the center. Compare the measurements with the manufacturer's specifications. If there are no specifications available, a rule of thumb is 0.003 in. (0.08 mm) for any six-inch length. Use the feeler gauge to determine whether the cylinder head needs to be resurfaced.

Use the straightedge in the same matter to check for warpage of the intake and exhaust manifold mating surfaces for warpage. The general rule for maximum warpage limit for the manifold mating surfaces allowed is 0.004 in. (0.1 mm).

Task B2 **Inspect and test valve springs for squareness, pressure, and free height comparison; replace as necessary.**

Begin valve spring inspection with a visual check for obvious signs of wear, cracks, corrosion, pitting, and nicks. If a valve has any of these defects, discard it and replace it with a new one. After a visual inspection, the spring must be checked for squareness, free length, and spring pressure.

If a spring is not square, it can side load the valve stem and the valve guide, causing excessive wear. The free length of the spring must be measured and compared with the specifications. Some springs that have a free length below specification can be corrected with the use of shims. To check a spring for squareness and free length, place the spring next to a square and rotate it while watching for warpage. Valve spring pressure is tested using a special spring tension gauge. The tension gauge works by compressing the spring to the specified height and observing the spring pressure reading on the gauge. Spring tension should fall within 10 percent of the manufacturer's specification, and no more than 10 pounds difference between springs.

Task B3 **Inspect and replace valve spring retainers, rotators, and locks; replace valve stem seals.**

Valve spring retainers and locks must be checked for wear and scoring. Check for signs of cracks and areas of discoloration. When any of these conditions are present, replace the components. The valve lock grooves on the valve stems must be inspected for wear, particularly round shoulders. If these shoulders are uneven or rounded, replace the valve. It is a good practice to replace the valve stem seals any time the cylinder head is disassembled.

Task B4 Inspect valve guides for wear, and check valve guide to stem fit; determine needed repair.

Before attempting, to measure valve guide wear and to measure valve seat run out, it is important to use a valve guide cleaner or a bore brush. After preparing the guide, use a small bore gauge and measure the valve guide at three different locations from top to bottom. Measure the fingers at each different location of the bore gauge with an outside micrometer. Measure the diameter of the valve stem, and subtract the difference to find the clearance. Another method is the valve rock method. Insert the correct valve into the guide and install the special tool that maintains the height of the valve during the inspection. Place a dial indicator with bracketry on the cylinder head with the tip of the dial indicator at a right angle to the valve head. Zero the dial indicator. Rock the valve and observe the clearance indicated on the dial.

The maximum amount of wear that most manufacturers recommend is 0.005 in. (0.12 mm) or less. The desirable clearances are 0.001 to 0.03 in. (0.025 to 0.080 mm) for intake and 0.0015 to 0.0035 in. (0.04 to 0.009 mm) for the exhaust valves. If any of the valve guides are out of specification, replace the guide.

Task B5 Inspect, replace, and/or grind valves.

There are two valve areas that require reconditioning: the stem tip and, more commonly, the valve face. If a stone is to be used, it must be dressed before any grinding is performed. Usually, after conditioning eight to ten valves, it is time to dress the stone again. Watch the valve face surface while grinding. If the surface is rough or the valve chatters during the process, the stone needs to be dressed. Refer to the valve grinding manufacturer's procedure. Refer to the following tips when resurfacing the valve.

- Locate the valve into the bottom of the chuck each time a new one is inserted.
- Use a full stroke, using the entire width of the stone.
- Keep cool oil flowing over the valve face.
- Never remove more material from the valve than is needed to provide a fresh surface.
- An interference fit of one degree is required for installation. You accomplish this by cutting the valve seat one degree different from the valve face.

Task B6 Inspect and grind valve seats and/or determine need for replacement.

Valve seats are usually ground at three different angles: the seat angle, the topping angle, and the throat angle. Any valve seat that is excessively worn or deeply scared must be replaced with a new one. Refer to the manufacturer's specifications for proper grinding angles. The following are some considerations when grinding valve seats:

- The seat contact pattern should be 1/16 to 3/32 in. (1.60 to 2.39 mm) wide.
- Dress each stone with a diamond bit-dressing tool before cutting any seat.
- Worn valve guides must be replaced before attempting to grind any seats.
- When working the grinding stone, lift it on and off the seat at a rate of 120 times per minute. This will prevent pressure on the seat and stone.
- When grinding the seat, only remove enough material to provide a new surface.
- Never apply pressure to the grinding stone because it will remove too much material and cause the stone to be dressed sooner.
- It is critical that the stone is dressed to the proper angle.
- Be sure to use the correct size pilot as a guide for the grinding seat assembly.

Task B7 Check valve face to seat contact and valve seat run out; service seats and valves as necessary.

When a valve is closed, it must make a pressure-tight seal. The valve face to valve seat contact provides the seal. In addition, this contact provides a path for heat to dissipate to the cylinder head.

First, measure the width of the seat with a machinist's ruler. If out of specification, remove material from the seat accordingly to obtain proper seat width.

Proper seat concentricity is critical for a proper seal. Use a valve run out gauge with an arbor installed in the valve guide bore. Slowly rotate the gauge around the seat while observing the readings. Compare the results with specifications; generally 0.002 in. (0.050 mm) is a usual tolerance. Remember, a worn valve guide will give an improper measurement.

Task B8 Check valve seat assembled height and valve stem height; service valve spring assembles as necessary.

The cylinder head is now ready for assembly. Make sure that the cylinder head has been cleaned. Before installing the valves, polish the valve stems with fine crocus cloth and solvent. Lubricate the valve stems with assembly grease or, at the least, engine oil. Bottom the valve against the seat and assemble the valve spring retainer with the locks onto the valve. While holding the valve seat, use a machinist's ruler and measure from the spring seat to the underside of the spring retainer. Compare this measurement to the specification and add shims, if necessary.

Stem height increases due to resurfacing the valve. Stem height is corrected by removing material from the valve tip. This measurement is from the valve tip to the spring seat. Again, use a machinist's ruler.

Task B9 Inspect pushrods, rocker arms, rocker arm pivots, and shafts for wear, bending, cracks, looseness, and blocked oil passages; repair or replace.

Before assembling the rocker arm and the related components, some inspection of the components should be performed. The rocker arms must be inspected for wear and damage. Rocker arms are constructed of cast iron, stamped steel, or aluminum. There are three areas on the rocker arm that receive high stress. These areas are the pushrod contact, the pivot, and the valve stem contact surface. The pushrod and the valve stem contact areas should be round and show signs of even wear. The oil passages in the rocker arms should be clean and free from debris.

Inspect rocker arm shafts (if applicable) for straightness and for excessive wear at the points where the rocker arm rides. The shafts should be free of scoring and galling. Inspect the shaft for wear at the places where the rocker arm shaft pivots. A slight polishing with no ridges indicates normal wear. Wear in this location usually indicates a lack of lubrication, which in turn generates excessive heat. The shaft itself can be checked for straightness by rolling it on a smooth surface.

Pushrods should be inspected for signs of wear and bending. Inspect the tips for normal wear patterns. Check the pushrod run out by rolling it on a known true surface; a piece of glass works well. Run out should not exceed 0.003 in. (0.08 mm).

Task B10 Inspect and replace hydraulic or mechanical lifters.

When inspecting valve lifters, the surface face of the lifter must be smooth with a centered circular wear pattern. The surface should be a convex counter-machined face. If wear extends to the edge of the lifter, the convex shape is worn away and the lifter must be replaced. The lifter body should be polished and smooth, free from any ridges, scaring, and signs of scuffing.

Hydraulic lifters that pass the visual inspection should be tested for leak down. This is done using a special tool that applies weight to a primed lifter submerged in engine oil. While the lifter is bleeding down, observe the scale and compare the rate of bleed down over a certain length of time. Leakdown range is between 20 and 90 seconds.

Sometimes it might be necessary to disassemble a lifter. Only disassemble one lifter at a time. Keep track of the order in which it comes apart.

Flat tappet lifters require a break-in period. A normal break-in involves applying break-in lube on the camshaft and the lifter face. Immediately following startup, the engine must be run at varying speeds between 1500 to 2500 rpm over a 20-minute period. The engine is run at off idle speeds to provide adequate oil splashed onto the machined surfaces that are in contact with the camshaft.

Task B11 Adjust valves.

Valve clearance usually needs to be adjusted after cylinder head reconditioning. The following are typical methods of adjusting valve lash:
- Adjustable nut attaching the rocker arm to the stud.
- Adjustable screw located in the end of the rocker.
- Selectable shims that are positioned between the camshaft lobes and the followers.
- Selective pushrod length.

Typical adjustments require that the number one cylinder be positioned at top dead center (TDC) on the compression stroke. Adjust the hydraulic lifters to the specified preload, while setting the specified clearance for solid lifter engines. After the number one cylinder is adjusted, the crankshaft must be rotated to the next designated location and the specified valve clearance adjusted. Depending on the manufacturer, all the valves can be adjusted at two different crankshaft locations, while others require locating each piston to TDC.

C. Engine Block Diagnosis and Repair (8 Questions)

Task C1 Inspect and replace pans, covers, gaskets, and seals.

Inspect the oil pan for cracks or dents. Also, inspect the pan for extensive areas of rust. If the oil pan were rusted, it would be the most convenient time to replace it. Check the gasket-mating surface of the pan for straightness, using a straightedge and a feeler gauge. If the gasket surface is warped, it can be straightened by striking it with a ball peen hammer on a flat, true surface. Stamped steel valve cover gasket mating surfaces can be straightened the same way that oil pan surfaces are aligned.

When installing new gaskets, make sure that the gasket and the surface that it is going to seal are free from debris and oil or grease. It is imperative that all of the old gasket material is removed for proper sealing.

Task C2 Inspect engine block for cracks, passage condition, core and gallery plug condition, and surface warpage; service block or determine repairs.

Cracks in the cylinder block are usually found during a visual inspection. When cracks are found, you should attempt to determine the cause. The following usually cause cracks:
- Fatigue
- Excessive flexing
- Impact damage
- Extreme temperature changes in a short period of time
- Detonation

Inspect all the oil passages with a shop light. The oil passages should be free from any gasket material, metal shavings, and other foreign objects. An engine brush kit can be used to clean all the oil passages.

Visually inspect the cylinder block deck for scoring, corrosion, cracks, and nicks. If a scratch in the deck is deep enough to catch on a fingernail, the deck needs to be resurfaced. Measure deck warpage with a precision straightedge and feeler gauge. To obtain proper results, the deck must be perfectly clean. Check for warpage across the four edges of the deck. The thickest feeler gauge that will fit between the straightedge and the deck determines the amount of warpage. If a deck is warped, the amount of material needed to be removed must be determined. On V-type engine blocks, when one deck is warped and needs to be machined, both sides have to be machined at the same time.

Task C3 **Inspect and measure cylinder walls for wear and damage; determine needed service.**

After visually inspecting the cylinder bores, use a dial bore gauge, an inside micrometer, or a telescoping gauge to measure the bore diameter. Piston movement in the cylinder bores produces uneven wear throughout the cylinder. The cylinder wears the most 90 degrees to the piston pin and in the area of the upper ring contact at TDC. This is because the cylinder receives less lubrication while being subjected to the greatest amount of pressure and heat at the top of the cylinder.

Taper in the cylinder bore causes the piston ring gaps to change as the piston travels in the bore. To measure taper and out-of-round using a dial bore gauge, simply rotate and move the gauge up and down in the bore. If you use a telescoping gauge or inside micrometer, measure the top of the bore just below the deck and at the **bottom of the ring travel.** As a general rule of thumb, the maximum taper allowed is 0.005 in. (0.125 mm) and 0.001 in. for out-of-round. If any cylinders are out of specification, all the cylinders should be bored to a standard oversize.

Task C4 **Remove cylinder wall ridges; hone and clean cylinder walls.**

The ridge at the top of the cylinder walls should be removed with a ridge reamer before attempting to remove the pistons. When using a ridge reamer, do not remove any more material than necessary. Begin the procedure by turning the crankshaft to lower the piston in the cylinder. Place an oiled rag on top of the piston to catch any metal shavings that might fall into the cylinder. Always follow the tool manufacturer's instructions for the ridge reamer that you are using. Rotate the tool in a clockwise rotation until the ridge is removed. Remove the rag from the cylinder and clean out any remaining metal shavings.

If the cylinder needs to be bored, it should be determined how much oversized it will need to be. Use the cylinder that has the worst wear as a reference for overbore size. It is a good practice to match the pistons to the bore. The pistons will have some differences in size, due to manufacturing tolerances. Measure the exact size of the replacement pistons. Then, determine the desired finished size of the cylinder and how much will be bored, leaving 0.003 to 0.005 in. (0.075 to 0.125 mm) for finishing by honing.

Task C5 **Inspect camshaft and bearings for wear and damage; determine bearing clearance and replace as necessary.**

Begin the camshaft inspection with a visual inspection of the lobes and journals. Both of these must be free of scoring and galling. Normal lobe wear pattern is slightly off center with a wider wear pattern at the nose than at the heel. The off center wear is a result of the slight taper of the lobe used in conjunction with the convex shape of the lifter to rotate the lifter. If the wear pattern extends to the edges of the lobe, the lifter will not rotate. If this condition exists, the camshaft must be replaced.

To measure camshaft bearing oil clearance, use an expandable bore gauge and a micrometer to measure the inner diameter of the bearing. Then use a micrometer to measure the expandable bore gauge. Subtract the two measurements to obtain the bearing oil clearance. Refer to the manufacturer's specifications for the proper oil clearance.

Task C6 **Inspect crankshaft for surface cracks and journal damage and wear; check oil passage condition; determine needed repair.**

The crankshaft transmits the torque from the connecting rods to the drivetrain. These pressures and rotational forces eventually cause wear and stress on the crankshaft. Before reusing the crankshaft, a thorough visual inspection of the subject areas is required. These areas include the main bearing journals, connecting rod journals, fillets, thrust surface, oil passages, counterweights, seal surfaces, and vibration damper journal. Next, inspect for warpage and measure the journals for excessive wear. An additional check is to inspect for stress cracks using magnetic particle inspection (MPI).

First, visually inspect the crankshaft for obvious wear and damage. This includes inspecting the threads at the front of the crank, the keyways, and the pilot bushing bore. When inspecting the journals, run your fingernail across the surface to feel for nicks and scratches. If a journal is scored, it must be polished before an accurate measurement can be obtained. Remember to inspect the area around the fillet very closely. Stress cracks can develop in this area.

Clean the oil passages by running a length of wire or a small bore brush through them, followed by a spray cleaner. Inspect the passage openings.

Task C7 — Inspect and replace main and rod bearings; check assembled clearances.

Inspect the main and rod bearings for wear patterns and record your determination in your notebook or on the work order. Note any unusual wear patterns indicating crankshaft or crankcase misalignment, lack of oil, and so forth. Inspect the backside of the bearings to determine if the crankshaft has been machined. Most oversized bearings are stamped to indicate the over- or undersize.

When an engine is reconditioned, the main and rod bearings are replaced. However, inspection of the old bearings provides clues as to the cause of an engine failure. For example, the soft material used to construct the bearings allows impurities to embed into it. Excessive metal flakes may alert the technician that there is metal-to-metal contact between moving parts in the engine.

To determine oil clearance with rod or main bearings use Plastigage®. Place the bearing inserts into the cap and in the saddle. Next, install the crankshaft into the cylinder block. The crankshaft must be free from oil so that the Plastigage® will stay in position while the cap is torqued. Select the proper size Plastigage® for the proper oil clearance and use a strip long enough to fit across the journal. Now, coat the bearing insert with oil to ensure that the Plastigage® will stick to the journal instead of the bearing. Install and torque the cap. Remove the cap and measure the width of the Plastigage® with the scale on the package. Use an oiled rag to remove all traces of the Plastigage® from the journal. After all clearances have been checked, remove the crankshaft and lubricate both the bearings and the crankshaft journals. If all clearances are within specifications, reinstall the crankshaft, caps, and the bolts.

Task C8 — Recognize piston and bearing wear patterns that indicate connecting rod alignment and bearing bore problems; inspect rod alignment and bore condition.

Before a connecting rod is accepted for reuse, it must be carefully inspected and measured. Both the big end and the small end bores should be inspected for clearance, out-of-round, and taper. In addition, the rod should be checked for center-to-length.

When measuring the inside diameter of the big end, assemble the cap to the rod, leaving the nuts loose. Place the rod in a soft jaw vise, with the jaws covering the parting line, and torque the cap nuts to specifications. This procedure ensures that the cap and rod are properly aligned with each other. The easiest way to measure the bore is to use a dial bore gauge. If an inside micrometer is used, measure the diameter in two or three directions near each end of the bore to obtain out-of-round and taper measurements. The greatest amount of out-of-round will occur in the cap and in a vertical direction.

Inspect the small end bore in the same manner as the big end bore. Use a dial bore gauge to measure clearance, out-of-round, and taper. If the piston assembly is designed with a press-fit pin in the connecting rod, the bore must be the correct size to provide interference fit. In addition, any scuffing or nicks in the bore may inhibit the piston from rocking properly. Sometimes the bore can be honed to a larger size to accept an oversized piston pin.

An Overview of the System Gasoline Engines (Test T1)

Task C9 Inspect, measure, service, or replace pistons, pins, and bushings.

The piston pin and the boss are subjected to severe operating conditions. Compounding this is the difference in expansion rates between the steel piston pin and the aluminum piston. A steel pin expands about 0.0003 in. (0.008 mm) for every 50° F increase in temperature. Consequently, proper clearance is critical.

Manufacturers vary concerning recommended procedures for checking pin clearance and how much clearance is allowed. For example, Chevrolet suggests measuring the pin diameter and the bore diameter; any clearance over 0.001 in. (0.025 mm) requires replacement of the pin and piston. Pontiac says the pin will fall through the bore with 0.0005 in. (0.01 mm) clearance, but not with 0.0003 in. (0.008 mm). Some import manufacturers check pin and bore wear by holding the piston and attempting to move the connecting rod up and down. Any movement felt indicates wear. Feel is not always the most reliable method. If you have any doubts, measure the clearance.

Measure the pin using the recommended procedure for the engine being serviced. With the pin removed, visually inspect it for wear. Use your hands to feel for scuffing or scoring. If there is any wear, the pin must be replaced.

Task C10 Measure piston-to-cylinder bore clearance; determine needed service.

Overheating of the piston generally causes seized or scored piston pins. Poor cooling system operation and improper combustion due to preignition or detonation can cause this. These overheating conditions cause lubrication failures of the pin.

Piston clearance is determined by measuring the size of the piston skirt at the manufacturer's sizing point. This measurement is subtracted from the size of the cylinder bore. If the piston clearance is not within specifications, it may be necessary to bore the cylinder to accept an oversized piston.

Since most pistons are camshaft ground, it is important to measure the piston diameter at the specified location. Some manufacturers require measurements across the thrust surface of the skirt centerline of the piston pin. Others require measuring a specified distance from the bottom of the oil ring groove. Always refer to the appropriate service manual for the engine that you are servicing.

Task C11 Replace piston rings; check ring groove clearance and end gap.

Measure the ring groove for wear. Installing a new ring backward in the groove and using a feeler gauge to measure the clearance can do this. Check the groove at several locations around the piston. If the side clearance is excessive, replace the piston. Excessive side clearance can result in ring breakage.

The topmost ring groove wears the most. Normal ring-to-groove side clearance is between 0.002 and 0.004 in. (0.05 to 0.10 mm). With a new ring located in the groove, attempt to slide in a feeler gauge the size of the maximum clearance specification. If the feeler gauge slides, the clearance is excessive. Locating the new ring into the ring groove in the same manner as checking clearance can also check the depth of the groove. Roll the ring around the entire groove while observing for binding. The ring depth should remain consistent. To check ring gap, place the ring into the appropriate cylinder bore. Use the piston to slide the ring to the specified depth, usually the bottom of ring travel in the stroke. The piston head will keep the ring square in the bore. Measure the gap at the ring ends. The value will be in the range of 0.004 in. per inch of diameter.

Task C12 Inspect, repair, or replace crankshaft vibration dampener (harmonic balancer) and flywheel.

Inspect the harmonic balancer for signs of wear in its center bore. Also, inspect the rubber mounting for indications of twisting and deterioration. If a balancer has slipped or rotated, it must be replaced. If wear to the center bore is present in the form of a groove worn in from the front engine seal, a repair sleeve kit is available.

Inspect the flywheel for signs of stress or heat cracks. Flexplates usually crack around the mounting hole area. Overheating because of a slipping clutch can cause these cracks. If the cracks are deep or the flywheel is blue in color, replace the flywheel.

Flywheels with light scoring and small cracks can be resurfaced before reuse. During resurfacing, flywheel warpage will also be eliminated. Flywheels and flexplates also contain the ring gear that the starter drive engages. Inspect the ring gear for missing or cracked teeth. If a flexplate ring gear is damaged, replace the flexplate. Most flywheels have replaceable ring gears.

D. Lubrication and Cooling Systems Diagnosis and Repair (8 Questions)

Task D1   Perform oil pressure tests; determine needed repairs.

When low oil pressure is evident, first check the oil level. Too low of an oil level will cause the oil pump to aerate and lose volume. If the oil level is too high, it may be caused by gasoline entering the crankcase because of a faulty fuel pump, improperly adjusted carburetor, sticking choke plate, or leaking fuel injector. If the level and the condition are not in question, use a mechanical gauge to check oil pressure.

When performing oil pressure tests, remove the sending unit from the engine. Use the appropriate adapters and connect the gauge to the engine. Start the engine and observe the pressure at idle. Watch the gauge while the engine warms up and note any pressure loss due to the temperature increase. Increase the engine speed to 2000 rpm while observing the gauge. Compare the test results with the manufacturer's specifications.

No oil pressure at all indicates that there is a problem with the oil pump driving mechanism or the oil pump itself. Other possible causes include:

- Oil pump pickup is plugged
- Gallery plugs are leaking
- A hole is present in the pickup tube
- Lower than specified oil level
- Improper oil viscosity
- Sticking or weak oil pressure relief valve
- Worn engine bearings

Task D2   Inspect, repair, or replace oil pump and drives.

The oil pump is usually replaced whenever the engine is rebuilt. If they are reused, they must first pass a visual inspection. If it fails an inspection, some manufacturers provide a rebuild kit. Proper inspection requires the oil pump to be cleaned and disassembled. Most often the pump is replaced rather than rebuilt because of the low cost of a new pump.

Oil pump drives can often be the cause of no oil pressure. Often the oil pump will ingest foreign particles, which are too large to pass through the gear assembly. This condition causes the pump to lock up, which will cause the drive shaft to break.

Task D3   Inspect, adjust, and replace belts and pulleys.

Since the friction surfaces are the sides of a V-belt, the belt must be replaced if the sides are worn and the belt is contacting the bottom of the pulley. The belt tension may be checked with the engine shut off, and a belt tension gauge placed over the belt at the center of the belt span. A loose or worn belt may cause a squealing noise when the engine is accelerated. Measuring the amount of belt deflection with the engine shut off also may check the belt tension. Use your thumb to depress the belt at the center of the belt span. If the belt tension is correct, the belt should have 1/2 in. deflection per foot of belt span.

An Overview of the System

Ribbed V-belts usually have a spring-loaded belt tensioner, with a belt wear indicator scale on the tensioner housing. If a power steering pump belt requires tightening, always pry on the pump ear, not on the housing.

Task D4 Perform cooling system pressure tests; determine needed repairs.

A pressure tester may be connected to the radiator filler neck to check for cooling system leaks. Operate the tester pump and apply 15 psi to the cooling system. Inspect the cooling system for external leaks with the system pressurized. If the gauge pressure drops more than specified by the vehicle manufacturer, the cooling system has a leak. If there are no visible external leaks, check the front floor mat for coolant dripping out of the heater core. When there are no external leaks, check the engine for combustion chamber leaks.

The radiator pressure cap may be tested with the pressure tester. When the tester pump is operated, the cap should hold the rated pressure. Always relieve the pressure before removing the tester.

Task D5 Inspect and replace thermostat, bypass, and housing.

If a customer brings in a vehicle with an overheating problem, it is possible that the thermostat is not opening. In addition, a thermostat that is stuck in the open position could cause an engine that fails to reach operating temperature. Check the temperature rating of the thermostat that is to be replaced, and confirm that it is the proper one for the engine application. Visually inspect the thermostat for rust and other contamination. Make sure that the thermostat was installed properly.

To test the thermostat, submerge it into a container of water. Use a thermometer so the temperature when the thermostat opens can be determined. Heat the water while observing the thermostat. At the rated temperature of the thermostat, it should begin to open.

Task D6 Drain, flush, refill, and bleed cooling system in accordance with accepted procedures.

Flushing of the cooling system is accomplished by using pressurized water through the cooling system in a reverse direction of normal coolant flow. A special flushing gun mixes low-pressure air with tap water. Reverse flushing causes the deposits to dislodge from the various components. They can then be removed from the system. The engine block and radiator should be flushed separately.

To flush the radiator, drain the system and disconnect the upper and lower hoses. Attach a long hose to the upper hose outlet to deflect the water. Disconnect and plug the heater hoses that are attached to the radiator. Fit the flush gun to the lower hose opening. This causes the radiator to be flushed in the reverse direction. Fill the radiator with water and turn on the gun in short bursts. Continue flushing until the water exiting the radiator is clean.

The cooling system is filled with a 50/50 mix of antifreeze to water. Before filling the cooling system, make sure that all hose clamps are tight and the drain plug is tight. Fill half of the system capacity with 100 percent antifreeze, the rest of the capacity with pure water. Continue to fill the cooling system as needed as the engine warms up.

Task D7 Inspect and replace water pump and hoses.

One replaces the cooling system water pump when the bearing has failed, or more commonly when the front lip seal has failed and has caused a coolant leak. Most water pumps can be removed after disconnecting the lower or upper radiator hoses at the pump. Remove any bypass or heater hoses that are attached to the pump. Remove the bolt attaching the pump to the front cover. When removing the bolts, take time to keep them in order since they are usually different lengths. The water pump might need a tap with a hammer to separate it from the cover. Remove all traces of gasket material from the front cover and be sure to use approved sealer with the new gasket.

Task D8 **Inspect and replace radiator, pressure cap, expansion tank, and coolant recovery system.**

The following steps can be used for removing a radiator from most vehicles.
1. Disconnect the negative battery cable.
2. Drain the cooling system.
3. Loosen the hose clamps and disconnect the hoses from the radiator.
4. If equipped, disconnect the transmission cooler lines.
5. If equipped, disconnect the electric fan connector.
6. Remove the bolts attaching the fan module or the shroud to the radiator.
7. If the air conditioning condenser cannot be disconnected from the radiator, the air conditioning system must be recovered.
8. Remove the upper radiator cross mount.
9. Remove the fan module or the shroud.
10. Remove the radiator.

Task D9 **Clean, inspect, and replace fan, fan clutch, and fan shroud.**

The cooling fan can be driven from an accessory belt or be electrically operated. Regardless of how it is powered, the fan blades must be inspected for stress cracks. Since the fan blades are balanced to prevent vibration to the water pump, if any are damaged the fan must be replaced. Belt driven fans use either flex fans or a viscous fan clutch. Flex fans should be inspected for stress cracks, while viscous fan clutches should be inspected for indications of leakage. Also, inspect the thermostatic spring for free movement and accumulations of dirt and debris. If any free movement exists, replace the clutch.

Electric fans are also inspected for damage and looseness. If the fan fails to turn on at the specific temperature, the temperature-sending unit is most likely the cause. If the fan is in question, the easiest method to check the operation is to apply direct battery voltage to the fan thermal. If the fan operates, it is functioning properly. If the fan does not operate, rule out the ground path first before faulting the motor.

E. Ignition System Diagnosis and Repair (11 Questions)

Task E1 **Diagnose no-starting, hard starting, engine missing, power loss, and/or poor mileage problems on trucks with point-type and electronic ignition systems; determine needed repairs.**

When diagnosing a no-start condition, many components could be the cause. For this instance, we will assume that the problem is with the ignition system. The ignition defects that may cause a no-start condition could be some of the following:
- Defective coil
- Defective cap and rotor
- Defective pickup coil
- Open secondary coil wire
- No primary voltage at the coil
- Fouled spark plugs
- Burned ignition points

If an engine has misfiring problems, the following components could be the cause:
- Engine compression
- Intake manifold leaks
- High resistance in the spark plug wires or coil secondary wire

- Electrical leakage in the distributor cap
- Defective coil
- Defective spark plugs
- Low primary voltage and current
- Improperly routed spark plug wires
- Insufficient dwell angle
- Worn distributor bushings

If a power loss condition exists, check the following:

- Engine compression
- Restricted exhaust or air intake
- Late ignition timing
- Insufficient ignition advance
- Cylinder misfire

The following can cause engine detonation:

- Higher than specified engine compression
- Ignition timing too far advanced
- Excessive centrifugal of vacuum advance
- Spark plug heat range too hot
- Improperly routed spark plug wires

Task E2   Inspect, test, repair, or replace ignition primary circuit wiring and components.

To diagnose a no-start condition with the primary ignition circuit, the following procedure can be used. Connect a 12-volt test lamp from the coil tachometer terminal to ground, and turn on the ignition switch. On high-energy ignition systems (HEI), the test light should be on because the module primary circuit is open. If the test light is off, there is an open in the circuit in the coil primary winding or in the circuit from the ignition switch to the coil battery terminal. On Chrysler electronic or Ford dura-spark II systems, the test light should be off because the module primary circuit is closed. Since there is primary current flow, most of the voltage is dropped across the primary coil winding. This action results in low voltage at the tach terminal, which does not illuminate the test light. On these systems, if the test light is illuminated, there is open in the module or in the wire between the coil and the module.

Crank the engine and observe the test light. If the light flutters while the engine is cranked, the pickup coil signal and the module are satisfactory. When the test lamp does not flutter, one of these components is defective. The pickup coil can be tested with an ohmmeter. If the pickup is satisfactory, the module is defective.

Task E3   Inspect, test, and service distributor.

When inspecting a distributor, start with a visual inspection. Check for signs of cracks on the cap, damaged wires, or terminals. Remove the cap and inspect the rotor and the inside of the cap for signs of arcing. Inspect all lead wires for worn insulation and loose terminals. Check the centrifugal advance mechanism for wear, particularly the weights for wear on the pivot holes. Also, inspect the pickup plate for wear and make sure that it moves freely. Test the vacuum advance unit using a hand-held vacuum pump. Apply 20 inches of vacuum to the advance unit and make sure that it moves the advance plate and holds vacuum. Inspect the drive gear for worn or chipped teeth. On point type ignitions, check the points for burning or pitting. Replace the points if either one of these signs are visible.

Task E4   Inspect, test, service, repair, or replace ignition system secondary circuit wiring and components.

If the primary ignition circuit is ruled out as a problem with a no-start condition, the secondary circuit must be inspected. Connect a test spark plug to a secondary ignition wire and ground the case. Be sure to use the proper spark plug for the vehicle application. Crank the engine and observe the plug. If the spark plug fires, it will indicate the coil is not the problem. Connect the spark plug to several different spark plug wires and crank the engine while observing the plug for each individual cylinder. If one or more plugs fire while others do not, this indicates that current is leaking through a defective distributor cap, rotor, or spark plug wires. If the test spark plug fires at all the individual wires, the system is functioning properly.

Task E5 Inspect, test, and replace ignition coils.

The ignition coil should be inspected for cracks and signs of electrical leakage. The cylinder type coil should be inspected for oil leaks. If oil is leaking from the coil, air space is present in the coil, which allows condensation to collect inside the coil. Condensation in an ignition coil causes high voltage leaks and engine misfiring.

Use an ohmmeter to test the primary coil windings. Attach the test leads to the terminals; an infinite reading indicates an open winding. The coil is shorted if the meter reading is below the specified resistance. Most primary windings have a resistance of 0.5 to 2 ohms.

Use an ohmmeter to test the secondary windings of the coil. Attach one test lead to the primary terminal and one to the secondary terminal. As before, an infinite reading indicates open in the circuit. Any reading below the specified resistance indicates a shorted secondary winding.

Task E6 Check and adjust ignition timing and timing advance/retard.

Ignition timing procedures vary from vehicle to vehicle. The procedure and the specifications for each vehicle are located on the underhood label. On distributors with advance mechanisms, manufacturers usually recommend disconnecting and plugging the vacuum advance hose while checking the timing. On carburetor engines, the engine speed must be set beforehand to the manufacturer's specifications. The timing light pickup is placed on the number one spark plug wire, and the power supply wires are connected to the battery terminals. The following steps can be used on most vehicles:

1. Connect the timing light, and start the engine.
2. Make sure the engine is idling at the specified rpm.
3. Aim the timing light at the timing tab next to the harmonic balancer and observe where the timing is set.
4. If the timing mark is not at the specified location, loosen the hold-down screw and rotate the distributor until the mark is at the specified location.
5. Tighten the hold-down screw and recheck the timing.
6. Connect the vacuum advance hose and any other components that were disconnected.

Many timing lights have the capability to check the spark advance. An advance control on the light slows the flashes of the light as the advance knob is rotated. When the light flashes are slowed with the engine running at higher speed, the timing marks move back to the basic setting.

Task E7 Inspect, test, and replace ignition system pickup sensor or triggering devices.

Distributor pickup gap can usually be adjusted. To adjust the gap, use a nonmagnetic feeler gauge positioned between the reluctor high points and the pickup coil. If a pickup gap adjustment is required, loosen the bolts, insert the correct feeler gauge, close the gap, and tighten the mounting bolts. Pickup coils that are riveted to the advance plate do not require adjustment.

Connect an ohmmeter to the pickup coil leads to test the pickup coil for an open or shorted circuit. While the leads are still connected to the ohmmeter, pull on the leads that are connected to the pickup coil and watch for an erratic reading. This would indicate an intermittent open in the wires. Most pickup coils have a resistance of 150 to

900 ohms. If the pickup coil has an open circuit it would be indicated by an infinite meter reading, whereas a meter reading below specifications would indicate a shorted coil.

Task E8 **Inspect, test, and replace ignition control module.**

Each vehicle manufacturer has its own ignition module tester. These testers check the module's ability to switch the primary ignition circuit on and off. Follow the instructions of the test equipment manufacturer. The module leads are connected to the test equipment while the power supply connections for the module are connected to the battery.

Task E9 **Inspect, test, and replace electronic and vacuum-type governor assemblies.**

To diagnosis a vehicle with a no-start condition, first it must be determined if the problem is related to the fuel or the ignition systems. We will only discuss problems related to the fuel system. First, the basics must be checked; fuel level and fuel condition.

F. Fuel and Exhaust Systems Diagnosis and Repair (10 Questions)

Task F1 **Diagnose no-starting, hard starting, poor idle, flooding, hesitation, engine missing, power loss, poor mileage, and/or dieseling problems on trucks with carburetor and fuel-injection systems; determine needed repairs.**

For fuel-injected vehicles, it must be verified that the fuel pump is functioning, then, that there is pressure and volume at the rail. Next, the injectors must be inspected for operation. Twelve-volt power must be available to the fuel injector. The powertrain control module (PCM) controls the ground path for the injectors. If it is suspected that the ground path is faulty, further diagnosis with a scan tool should take place to determine if there are any fault codes or problems with the injector drivers located in the PCM.

The carburetor must be inspected before it is determined that it needs a rebuild. Check that the choke plate is functioning properly along with the accelerator linkage. If possible, check the float level. A faulty float level can cause flooding, as well as fuel starvation.

Task F2 **Inspect, test, and/or replace fuel pumps (mechanical and electrical); inspect, service, and replace fuel filter, gas cap, fuel lines, fuel tank, hoses, and fittings.**

Electric fuel pumps operate for five to ten seconds during ignition key up. The reason for this is to run the pump so that the fuel rail can be pressurized at startup. The PCM controls the relay that supplies power to the electric fuel pump.

A cam that is located on front end of the camshaft drives most mechanical fuel pumps. These fuel pumps are bolted to the engine block. The pump must be observed for signs of external or internal leakage. Internal leakage will usually contaminate the engine oil with raw fuel. This can be a dangerous situation if not attended to. Mechanical fuel pumps have two main lines, one from the fuel tank, and one to the carburetor.

Task F3 **Remove and replace carburetor, adjust linkages.**

To remove the carburetor, all linkages, vacuum, and fuel lines must be disconnected. Take your time and, if necessary, label all the connections so that reassembly will not be confusing. After disconnecting the vacuum and linkage connections, loosen the fuel supply line with a flare nut wrench. Loosen the fuel line fitting one or two turns only. Now, remove the retaining hardware that holds the carburetor to the intake manifold. After the bolt or nuts have been removed, disconnect the fuel line from the carburetor. Take the proper cautions, since there will be liquid fuel lost from the carburetor and the fuel supply line.

Task F4 **Rebuild carburetor including: disassembling, cleaning, replacing faulty parts, assembling, and adjusting.**

After removing the carburetor, remove any plastic parts that can be dissolved in cleaning solution. If the carburetor has a large amount of varnish and grime, it may be soaked in cleaning solution before disassembly, otherwise disassemble the carburetor and soak the parts in a cleaning solution for half an hour to an hour. During disassembly, check for obvious signs of damage or worn-out parts. Check springs for signs of binding or broken coils. Inspect seals for tears or signs of swelling. Observe where all the check balls are placed so that they can be replaced. Inspect all the metering passages for obstructions. Check the subassemblies for proper operation and broken parts. The subassemblies include the choke, float and linkage, dashpots, and accelerator pump.

Task F5 **Remove, clean, and replace throttle body; adjust related linkages.**

To remove the throttle body, first remove the air intake assembly that is connected to the throttle body. Disconnect the throttle linkage and if equipped, the transmission kickdown cable. Disconnect the throttle position sensor and remove the throttle body. Remove the old gasket, since a new one will be installed during installation. Be sure to place a rag in the intake manifold opening to ensure that no foreign objects enter the engine. Some throttle bodies have a special coating that does not allow carbon to build up on the surface of the throttle plate. If the throttle body has a special coating, no chemical cleaners can be used to clean the throttle body.

Task F6 **Inspect and replace carburetor/fuel injection mounting plates, intake manifold, and gaskets.**

Faulty intake manifold gaskets can cause driveability problems that can commonly be confused with other fuel system components. A defective gasket in the intake system can easily be overlooked. Perform specific tests to determine if any intake gaskets have failed. It is a good practice when replacing the intake manifold gaskets to replace all the gaskets in the intake system. Conveniently, they usually come together in a kit. Be sure that the gasket(s) that you are replacing are made from the same material as the original equipment manufacturer (OEM).

Task F7 **Inspect, clean, and adjust carburetor choke (automatic and manual).**

The function of the automatic choke is to provide the best possible engine performance in cold temperatures. A proper operating choke should open and close freely. The purpose of the choke is to open as soon as the engine can run without its help. Automatic chokes that use a thermostatic coil have index markings on the plastic cover. The index marks indicate richer or leaner in adjustment. Check for proper operation of the choke plate before attempting to adjust the choke. Make sure that the choke linkage is free from dirt and debris. In addition, the choke spring must be "cold" before attempting an adjustment. This adjustment is nothing more than the amount of air that passes the choke plate when the choke is fully applied.

Task F8 **Inspect, test, adjust, and replace cold enrichment systems components.**

The fuel-injection system has to enrich the fuel to air mixture during cold temperatures. This is accomplished by means of electronically controlling the dwell or the pulse width of the injector. The pulse width of the injector is the amount of time that the injector is open. This time is rated in milliseconds. The PCM monitors the engine coolant temperature sensor; if the engine temperature is below a specified level, the strategy is programmed to increase the pulse width. Engine control strategies monitor (in order of importance) the engine coolant sensor, the oxygen sensor (H_2O), and the mass airflow sensor.

When the engine coolant reaches a specified temperature, anywhere from 130° to 170°F (54° to 76°C) and the oxygen sensor reaches 600°F (315°C), the system enters closed loop. In this mode, the fuel mixture is adjusted according to the oxygen sensor, which measures the content of oxygen that is left over after the combustion process.

Task F9 Inspect, test, and replace deceleration fuel reduction or shut off systems components.

The PCM determines deceleration by observing the following sensors (in order of importance):

- Throttle position sensor
- Engine vacuum
- Engine speed
- Vehicle speed

When the driver releases the accelerator pedal, the throttle position sensor value immediately changes, indicating to the PCM a change has occurred and the driver wants to decelerate. Next, the engine vacuum will increase as the throttle plates close. This is the second indication of the driver's desire to decelerate and is measured by the manifold absolute pressure (MAP) sensor. Now that the driver has made a change and enough time has passed for the PCM to make a change, the engine will decrease in speed. The result is the vehicle will decrease in speed.

Task F10 Inspect, service, and repair or replace air filtration system components.

The automotive engine burns about 9,000 gallons (34,065 liters) of air for every gallon of gasoline at an air-fuel ratio to 14.7 to 1. With today's engines, which run much leaner to meet emissions standards, the ratio is closer to 10,000 gallons (37,850 liters) of air. This is a large amount of air that needs to be filtered. This is accomplished by using a paper element that is enclosed in a sealed box that is attached to the throttle body through intake plumbing. The paper element is disposable and should be replaced at intervals of 10,000 to 20,000 miles. A good practice to follow is to remove the air cleaner element and hold a shop light directly behind it while trying to see any light that passes through the element. If no light passes through the element, the filter should be replaced.

Task F11 Remove, clean, inspect/test, and repair or replace fuel system vacuum and electrical components and connections.

The fuel pressure regulator controls the amount of fuel that is returned to the fuel tank. This is also how the proper fuel pressure is maintained in the fuel rail. The fuel pressure regulator accomplishes this by sensing the engine manifold vacuum. As engine manifold vacuum decreases, which indicates the engine is under a load, the spring in the regulator closes the orifice and raises the fuel pressure. The higher fuel pressure is needed to enrich the fuel mixture under load.

The fuel pressure regulator is bolted to the fuel rail near the fuel return line connection. To remove the fuel pressure regulator, first relieve the fuel rail pressure to prevent fuel from being sprayed in the engine compartment. Always replace the old regulator gasket with a new one.

Task F12 Inspect exhaust manifold, exhaust pipe, muffler, and other components of the exhaust system; repair/replace as needed.

Exhaust manifolds are made of cast iron, which offers good resistance to change from temperature. The exhaust manifold bolts to the cylinder head. All the corrosive exhaust gas collects in the manifold and then is routed to the catalytic converter and out through the muffler and tailpipe. Vehicle manufacturers now use stainless steel for exhaust manifolds. Stainless steel offers greater resistance to corrosion and quicker warm-up, which helps the catalytic converter reach operating temperature sooner. Stainless steel is considerably lighter than cast iron.

Exhaust manifolds should be inspected for cracks and for extreme corrosion that can cause exhaust leakage. Exhaust manifold gaskets can sometimes become deteriorated and blownout from between the manifold and the cylinder head.

G. Battery and Starting Systems Diagnosis and Repair (7 Questions)

Task G1 Inspect, clean, fill, or replace battery.

Inspect the battery for cracks and signs of electrolyte leakage. Check that the terminals are tight and not heavily corroded. Also, check to make sure that the battery hold-down is in place and holding the battery securely. Use a solution of one part baking soda and one part water to clean the battery. Pour the solution over the entire battery, allowing it to run down the side and onto the battery tray. After the solution has been used up, rinse the battery with a generous amount of water. Some batteries allow you to add water to the cells. Use only distilled water to top off the cells. When replacing the battery, always disconnect the negative battery cable first. This is to prevent a short if the wrench touches a ground while disconnecting the positive cable. When installing the battery, connect the positive battery first, for the same reason.

Task G2 Slow and fast charge a battery.

Slow charging a battery is charging the battery at 1.5 to 7 amps per hour. The slower a battery is charged, the more complete the charge. Fast charges are usually 8 to 10 amps per hour. A battery can be fast charged at any rate (within reason) as long as the electrolyte does not boil over and the temperature does not exceed 120°F (50°C).

Task G3 Perform a battery capacity (load, high rate discharge) test; determine needed service.

A battery load test is performed to determine the electrical storage capacity of the battery. The test is performed by applying a load in amp rating half that of the cold cranking amperage of the battery. For example, if a battery has a cold cranking rating of 800 amps, to load test apply a load of 400 amps for 15 seconds. The battery voltage should not drop below 9.6 volts for the 15 seconds of the load test.

Task G4 Jump start a vehicle with jumper cables and a booster battery or auxiliary power supply in accordance with manufacture's specifications.

The accessories must be off in both vehicles during the boost procedure. The negative booster cable must be connected to an engine ground in the vehicle being boosted. Always connect the positive booster cable, followed by the negative booster cable, and complete the negative cable connection last on the vehicle being boosted. When disconnecting the booster cables, remove the negative booster cable first on the vehicle being boosted.

Task G5 Inspect, clean, repair, and replace battery cables and clamps.

After removing the battery cable from the post, clean the post and the terminals with a wire brush. It is always wise to spray the cable clamps with a protective coating to protect corrosion. A little grease or petroleum jelly will also prevent corrosion. In addition, protective pads are available that go under the clamp and around the terminal to prevent corrosion.

An Overview of the System   Gasoline Engines (Test T1)   51

Task G6 **Inspect, test, and replace starter relays and solenoids.**

Conditions of high current draw, low cranking speed, and low cranking voltage usually indicate a defective starter. This condition also may be caused by internal engine problems, such as partially seized bearings. Low current draw, low cranking speed, and high cranking voltage indicate excessive resistance in the starter circuit. A defective starter relay or solenoid can also cause this resistance. Relays and solenoids have disks that transmit the current; these disks become pitted and burned. This is what causes high voltage drops.

Task G7 **Remove and replace the starter.**

When reinstalling the starter motor you should perform a free spin test or a current draw test. You should also test the pinion gear clearance. Do this by disconnecting the M-terminal so you will be able to shift the pinion gear into the cranking position. Then you will be able to check the clearance with a feeler gauge. Normally, specifications call for a clearance of 0.010 to 0.140 in. (0.25 to 0.35 mm).

H. Emissions Control Systems Diagnosis and Repair (7 Questions)

Task H1 **Test, inspect, clean, and replace exhaust gas recirculation (EGR) valves, valve manifolds, and passages of the EGR system.**

Exhaust gas recirculation (EGR) valves redirect a metered amount of exhaust gas into the intake manifold. The reason for this is to help reduce exhaust emissions. Most EGR valves are vacuum operated. Some are controlled by means of an electric actuator. To test an EGR to tell if it is functioning properly, use a hand operated vacuum pump and apply 10 to 15 inches of vacuum to the port on the EGR. Within this range, the valve should open fully. When an EGR valve is stuck in the open position, the engine will run rough at idle and experience power loss and poor fuel economy. Depending on the application, the EGR passages in the manifold will become blocked and will need to be cleaned. Using special fuel-injection spray cleaner, or any cleaner that is safe to use with vehicles with oxygen sensors, can accomplish this. Spray the cleaner into the intake manifold passages to dissolve the carbon.

Task H2 **Inspect, repair, and replace controls and hoses of the EGR system.**

The EGR controls include the following:
- Coolant temperature override switch
- EGR Pressure Transducer (EPT)
- EGR Vacuum Regulator (EVR) solenoid
- Vacuum hoses
- PCM

Inspect all the hoses for cracking or any areas that could cause a vacuum leak. Most EPT and EVR solenoids are made of plastic and should be inspected where the vacuum lines attach.

Task H3 **Test the operation of the air-injection system.**

To check the operation of the air-injection system, your inspection should include the following:
- Check the condition of the drive belt and the tension on the belt.
- Inspect all the air system hoses and vacuum lines. Look for signs of cracking, brittleness, or burning.
- Inspect the check valves for exhaust leakage.

- Disconnect the hose at the pump side of the check valve, start the engine, and check for airflow from the hose. There should be adequate airflow at the hose opening. Increase the engine speed. As the engine speed increases, so should the airflow. Next, pinch off the pump outlet hose. A popping noise should be heard as the pressure valve opens.

Task H4 — Inspect, repair, and replace pumps, pressure relief valves, filters, pulleys, and belts of the air-injection system.

Air pumps do not require periodic maintenance, and one cannot rebuild an EGR valve. When removing or servicing the air pump, never pry against the pump housing, since it is made of aluminum and can be cracked easily. Removing the air pump is similar to removing an alternator or any engine-driven accessory. The following steps are for general air pump removal.

- Disconnect the vacuum and output hoses from the pump.
- Loosen the pump mounting bolts and remove the belt. Depending on the vehicle application, some of the other belts may have to be removed.
- Remove the pump bolts and remove the pump. If necessary, remove the pulley and transfer it to the new pump.
- Remove the diverter valve from the old pump and transfer it to the new pump.

Task H5 — Inspect and replace bypass (anti-backfire) valves and vacuum hoses of air-injection systems.

The main purpose of the bypass valve is to redirect the air to the intake manifold during deceleration to prevent backfire. Check the bypass valve as follows:

- Check the condition of all the hoses and the hose connections.
- Start the engine and chock the wheels.
- Disconnect the vacuum line at the bypass valve and place a finger over the end of the hose. If there is no suction, the hose is plugged or broken.
- Reconnect the vacuum hose and increase the engine speed to 200 rpm, then quickly close the throttle. If the engine backfires, the valve is defective and needs replacement.

Task H6 — Inspect, service, and replace hoses, check valves, air manifolds, and injectors of the air-injection systems.

The air-injection hoses should be inspected for cracks, deterioration, and wear. Check the hoses for air leaks. Any hoses that are replaced must be the same preformed shape as the original. Any hose that is not suited for air-injection systems may deteriorate prematurely. Most manifolds are one-piece assemblies and are connected to the exhaust manifold with nozzle fittings. Before attempting to loosen the nuts at the manifold, apply penetrating oil to the nuts so that the nuts can be loosened with fewer or no problems. Use a tube wrench to loosen the nuts. If the nuts still are a problem, heat from a torch can be used to persuade the nut. When installing the new air-injection manifold, use anti-seize compound on the threads to help start the threads and to prevent rusting between the threads.

Task H7 — Inspect, test, service, and replace positive crankcase ventilation (PCV) valve and system components.

The positive crankcase ventilation (PVC) system requires service at regular intervals to ensure that the system operates properly. Sludge and carbon can plug the PCV valve. Service the PCV valve by inspecting it, making sure that the valve is not clogged and it moves freely. Some symptoms of a PCV system that is not functioning properly include the following:

- Increased oil consumption
- Diluted or dirty oil, or oil that has sludge or is acidic

- Excessive blowby, or oil level dipstick that is blown out of its seat
- Rough idling or stalling at idle or low engine speeds

Inspect all PCV system hoses for cracks and deterioration. Check the air filter for oil deposits and for oil puddling in the oil cleaner housing. In addition, the system should be inspected for clogs. Clogs are usually caused by not changing the oil at regular intervals and by not maintaining the PCV system. Regular maintenance should include cleaning the PCV valve with a carburetor or fuel-injector cleaner.

Task H8 Inspect, test, and replace components of the catalytic converter systems.

Typical catalytic converter replacement is essentially the same as replacing a muffler or a tailpipe. Always wait until the exhaust system has cooled sufficiently. Use heat from a torch and a hammer to persuade the converter from the converter inlet pipe. Use the correct application converter for the vehicle. Remember to install all the heat shields back onto the converter.

I. Computerized Engine Controls Diagnosis and Repair (6 Questions)

Task I1 Diagnose the causes of emission problems resulting from failure of computerized engine controls.

Depending on what type of sensor or computerized fuel-injection component has failed, the results can range from poor fuel economy to the system operating in failure or limp-in mode. When a sensor fails, the PCM will recognize there is a fault and will illuminate the check engine lamp. If the component that has failed is one that is critical to engine driveability, the PCM will substitute a known good value so that the engine can still operate. This is what is known as limp-in mode. For instance, with a faulty oxygen sensor sending information to the PCM, the PCM may never think the engine is at operating temperature. With the PCM thinking that the engine is always cold, it will always provide a rich fuel mixture. A rich fuel mixture can cause the catalytic converters to overheat and melt down and can cause the engine oil to become sludged or diluted with liquid fuel. Any type of computerized fuel injection that has failed should be attended to as soon as possible to prevent other problems.

Task I2 Perform analytic/diagnostic procedures on vehicles with on-board diagnostic computer systems; determine needed action.

Trouble codes and other diagnostic tests can be performed with a hand-held scan tester. Scan tester operation varies depending on the make of the tester, but the following is a typical example:

1. Warm the engine until the engine is at normal operating temperature. Turn the ignition off. Connect the power adapter to the cigar lighter, and connect the other lead of the tester to the data link connector (DLC).
2. Follow the procedure that the scan tool provides on the display. These include entering the model year, engine size, and vehicle model. These are all listed in the tester operation manual. Now that the initial entries have been selected, different entries appear. A typical list of the next choice of entry is:
   - Engine
   - Anti-lock brake system
   - Suspension
   - Transmission

The technician must make the selection of what system they want to troubleshoot. Next, they have to decide if they want to retrieve stored trouble codes or see a live data screen.

Task 13 Inspect, test, adjust, and replace sensor, control, and actuator components and circuits of computerized engine control systems.

A defective engine coolant temperature (ECT) sensor may cause the following problems:

- Hard starting
- Rich or lean air-fuel ratio
- Improper operation of emissions devices
- Reduced fuel economy
- Driveability problems
- Engine stalling

An ECT sensor can be diagnosed easily with a scan tool or a digital voltmeter. An ECT sensor can be removed and placed in a container of water that is heated. Connect an ohmmeter across the terminals, and place a thermometer in the container. When the water is heated, the sensor should have the specified resistance at any temperature. If the sensor does not have the proper readings, replace it.

Task 14 Use and interpret digital multimeter, digital volt ohmmeter (DMM, DVOM) readings.

The use of digital multimeters can be very helpful when diagnosing sensors or circuits of the fuel-injection system. Multimeters provide several different types of readings: DC volts, AC volts, ohms, amperes, and milliamperes. With an ECT sensor installed in the engine, the sensor may be back-probed to connect the multimeter (set on DC volt scale) to the sensor terminals. The sensor should provide the proper voltage drop at the given coolant temperature.

To diagnose a throttle position sensor (TPS) with the ignition switch on, connect the voltmeter to the 5-volt reference wire to ground. If there is no voltage to the reference wire, the problem is between the electrical connector and the PCM.

Task 15 Read and interpret technical literature (service publications and information).

Technical service manuals and other published literature, such as wiring diagrams and power train and emission control diagnostic books, can be very beneficial. One should not overlook these diagnostic tools. The vehicle manufacturers publish the material that is contained in the books. These books have instructions in the front of the book that are easy to use.

Appropriate service repair procedures are essential for the safe, reliable operation of all motor vehicles as well as the personal safety of the technician performing the work. The service manuals provide general directions for accomplishing repair work with tested techniques. There are numerous differences in procedures, techniques, tools, and the parts used to service the vehicle.

The skill of the technician plays a large role in the quality of the service done. A service manual cannot anticipate all types of situations and variations that will occur to the technician performing the work. Any technician that deviates from the written procedure provided in the service manual must understand that they are compromising personal safety and risking damage to the vehicle that is being serviced.

Task 16 Test, remove, inspect, clean, service, and repair or replace electrical distribution circuits and connections.

Most automotive circuits are protected from high current flow that would exceed the capacity of the circuit's conductors and/or loads. Excessive current results from a decrease in the circuit's resistance. When the circuit's current reaches a predetermined level, most circuit protection devices open and stop current flow in the circuit. This is what prevents damage to the circuit and its components.

The most commonly used circuit protection devices are the fuses. A fuse contains a metal strip that will melt when the current flowing through it exceeds its rating. The thickness of the metal strip depends on the amp rating of the fuse.

A vehicle may have one or more fusible links to provide protection for the main power wires before they are divided into smaller circuits.

A circuit that is susceptible to an overload on a routine basis is usually protected by a circuit breaker. Some circuit breakers require manual resetting by pressing a button. Others must be removed from the circuit to reset themselves. Some circuit breakers are self-resetting. This type of circuit breaker uses a bimetallic strip that reacts to excessive current. When an overload or a circuit defect occurs that causes an excessive amount of current draw, the current flowing through the bimetallic strip causes it to heat. As the strip heats, it bends and opens the contacts. Once the contacts are open, the current can no longer flow.

Task 17 Practice recommended precautions when handling static-sensitive devices.

When working with electronic components steps must be taken to prevent damage to the components by static discharge. The human body can generate voltages as great as 35,000 volts by simply walking across the carpet on a dry day. As little as 300 volts can severely damage the sensitive electronic components of the vehicle. Due to the sensitive nature of these components, steps must be taken to protect them from static discharge. The technician must be grounded to safely drain off any static charge, and the electronic parts should be placed on a grounded conductive mat, rather than the vehicle's carpet or upholstery. Any time there is contact with sensitive electronics the technician must remain grounded. Touching a metal object is not sufficient. If a replacement part is sensitive to static discharge, it will normally be shipped in a static bag. The part should not be removed from the bag until the technician is properly grounded and the part is ready to be installed.

Task 18 Diagnose driveability and emissions problems resulting from failures of interdependent systems, (engine alarm, air conditioning, and similar systems).

Some vehicles have separate computers for engine control, transmission control, ABS, traction control, suspension control, lubrication control, and A/C control. On some trucks, many of these computers interconnect to the fuel management or engine computer via data links or they are independent or discrete from the fuel management or engine computer. One or more of these computers use some of the engine input sensors connected by data links to these computers to transmit their data.

If the data links are defective, the necessary inputs will not reach the engine computer or the other systems, causing both engine faults and possible faults in the interdependent computers. Defective data links may cause malfunctioning of some other output. For example, the engine coolant temperature (ECT) sensor signal is sent from the powertrain control module (PCM) to the transmission control module (TCM) for proper lockup converter control. If the data links are defective and the ECT signal is not available to the TCM, the lockup converter may not function and will send a message identifier (MID) transmission code. Yet, the engine fuel management will function normally because the data link is the problem, not the sensor itself. The scan tool, diagnostic reader, or OEM software with a laptop computer is used to test these data links and system data display readouts.

Sample Test for Practice

Sample Test

Please note the letter and number in parentheses following each sample question. They match the overview in section 4 that discusses the relevant subject matter. You may want to refer to the overview using this cross-referencing key to help with questions posing problems for you.

1. Technician A says that customer complaints must be listened to closely. Technician B says that when diagnosing a vehicle problem, always start with the easiest test to perform. Who is right?
   A. A only
   B. B only
   C. Both A and B
   D. Neither A nor B   (A1)

2. A customer has brought in a vehicle that has an engine oil leak. Upon inspection, the lower half of the engine is covered in oil. Technician A says that the engine must be steam cleaned before an accurate diagnosis can be given. Technician B says that if the lower portion of the engine is leaking, it is most likely the oil pan gasket. Who is right?
   A. A only
   B. B only
   C. Both A and B
   D. Neither A nor B   (A2)

3. What can be determined by the color of the engine oil?
   A. The age of the engine
   B. If the engine oil needs to be changed.
   C. How many miles it has been since the last oil change.
   D. Main bearing wear.   (A3)

4. An engine has an engine knock when it is started. What could be the cause?
   A. Bent pushrods
   B. One or more collapsed lifters
   C. Lack of oil pressure to the valve train
   D. Worn main bearings   (A4)

5. Blue smoke that is coming out of the tailpipe can be caused by all of the following **EXCEPT:**
   A. worn valve guides.
   B. worn valve seals.
   C. fouled spark plugs.
   D. worn piston rings.   (A5)

6. With a vacuum gauge connected to an engine, referring to the figure above, the gauge readings are steady at 18 in. Hg at idle, and when the engine speed is raised to 1500 rpm, the reading raises to a steady 25 in. Hg. Technician A says that the engine has a blocked catalytic converter. Technician B says that the engine is functioning properly. Who is right?
   A. A only
   B. B only
   C. Both A and B
   D. Neither A nor B (A6)

7. A vehicle with a suspected intake manifold gasket leak is in the shop. Technician A says that an intake manifold leak can cause a misfire at low rpm or idle and can disappear at higher engine speeds. Technician B says that engines sometimes stumble at idle and that retorquing the manifold could solve the problem. Who is right?
   A. A only
   B. B only
   C. Both A and B
   D. Neither A nor B (A7)

8. Technician A says that if an engine has one cylinder that is oversized, it will have a higher compression reading than the others. Technician B says if the compression pressures raise after adding a small amount of oil to the cylinder, the valve guides are worn. Who is right?
   A. A only
   B. B only
   C. Both A and B
   D. Neither A nor B (A8)

9. When performing a cylinder leakage test, which of the following is considered to be the maximum tolerable percentage?
   A. 5 to 10 percent
   B. 30 to 40 percent
   C. 40 to 45 percent
   D. 15 to 20 percent (A9)

10. Technician A says that engine analyzers are obsolete and cannot be used on current ignition systems. Technician B says that engine analyzers can be used to diagnose secondary ignition circuits. Who is right?
    A. A only
    B. B only
    C. Both A and B
    D. Neither A nor B (A10)

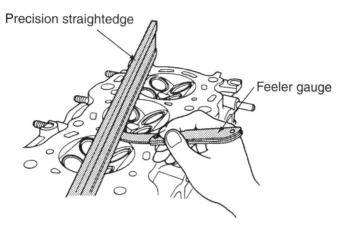

11. In the figure above, when measuring the cylinder head for warpage with a straightedge, the feeler gauge measurement is 0.025 in. (0.63 mm). Technician A says that the cylinder head needs resurfacing. Technician B says that the measurement is within tolerance. Who is right?
    A. A only
    B. B only
    C. Both A and B
    D. Neither A nor B   (B1)

12. Engine valve springs should be inspected for all of the following **EXCEPT:**
    A. cracks.
    B. spring tension.
    C. pitting and nicks.
    D. spring coil gap.   (B2)

13. Technician A says to inspect the valve spring retainers for cracks and discoloration. Technician B says that it is a good practice to replace the valve spring retainers when the valves are serviced. Who is right?
    A. A only
    B. B only
    C. Both A and B
    D. Neither A nor B   (B3)

14. What preparation must be done to the valve guides before measuring valve seat run out?
    A. Use 30 weight engine oil to lubricate the guide.
    B. Clean the valve guide with a bore brush.
    C. Ream the valve guide to its original diameter.
    D. Use a pilot to hold the guide in place.   (B4)

15. When resurfacing valves, Technician A says that valve stems usually do not require resurfacing. Technician B says that there is no need for an interference angle between the valve and the seat. Who is right?
    A. A only
    B. B only
    C. Both A and B
    D. Neither A nor B   (B5)

16. All of the following are considerations when grinding valve seats **EXCEPT:**
    A. dress the stone before cutting any seat.
    B. remove only enough material to provide a new surface.
    C. use transmission fluid to lubricate the stone only.
    D. do not apply pressure to the grinding stone.   (B6)

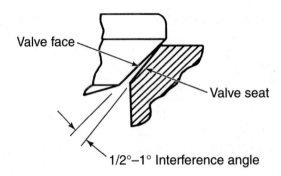

1/2°–1° Interference angle

17. Referring to the figure above, Technician A says that proper valve seat concentricity is critical for a proper seal, and a usual tolerance for seat run out is 0.002 in. (.50 mm). Technician B says that not only does the valve to seat contact provide an airtight seal but also a path for heat to dissipate. Who is right?
    A. A only
    B. B only
    C. Both A and B
    D. Neither A nor B (B7)

18. How is valve stem height corrected?
    A. By removing material from the valve tip
    B. Adding shims
    C. Installing a valve spring that is the correct height
    D. Removing material from the valve seat (B8)

19. Which area on a rocker arm is LEAST likely to receive a large amount of stress?
    A. The pushrod contact
    B. The pivot
    C. The valve stem contact surface
    D. The main body (B9)

20. Which of the following types of wear patterns is most likely desired for the lifter mating surface?
    A. Convex counter-machined face
    B. Smooth
    C. Centered circular wear pattern
    D. Crosshatched (B10)

21. While adjusting mechanical lifters, Technician A says when the valve clearance is checked on a cylinder, that cylinder should be positioned at TDC on the exhaust stroke. Technician B says some mechanical lifters have removable shim pads available in various thickness to provide proper valve clearances. Who is right?
    A. A only
    B. B only
    C. Both A and B
    D. Neither A nor B (B11)

22. When inspecting an oil pan that is leaking, Technician A says that it could be a crack or a rusted hole in the pan. Technician B says that it could be the steel flange that mates to the engine block. Who is right?
    A. A only
    B. B only
    C. Both A and B
    D. Neither A nor B (C1)

23. Technician A says that a stamped steel oil pan that is leaking from the flange cannot be repaired and needs replacement. Technician B says that the flange can be straightened by striking the surface with a ball peen hammer on a true flat surface. Who is right?
    A. A only
    B. B only
    C. Both A and B
    D. Neither A nor B (C1)

24. Technician A says a warped cylinder head mounting surface on an engine block may cause valve seat distortion. Technician B says a warped cylinder head mounting surface on an engine block may cause coolant and combustion leaks. Who is right?
    A. A only
    B. B only
    C. Both A and B
    D. Neither A nor B (C2)

25. While discussing cylinder measurement, Technician A says the cylinder taper is the difference between the cylinder diameter at the top of the ring travel, compared to cylinder diameter at the center of the ring travel. Technician B says cylinder out-of-round is the difference between axial cylinder bore diameter at the top of the ring travel, compared to the thrust cylinder bore diameter at the bottom of the ring travel. Who is right?
    A. A only
    B. B only
    C. Both A and B
    D. Neither A nor B (C3)

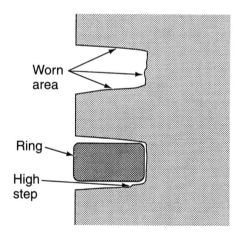

26. Referring to the figure above, if new rings are installed without removing the piston ring ridge, which of the following may result?
    A. The piston skirt may be damaged.
    B. The piston pin may be broken.
    C. The connecting rod bearings may be damaged.
    D. The piston ring lands may be broken. (C4)

27. When inspecting a used camshaft, Technician A says that a normal wear pattern will have a slightly off center with wider wear pattern at the nose than at the heel. Technician B says that the wear pattern should extend to the edge of the lobe. Who is right?
    A. A only
    B. B only
    C. Both A and B
    D. Neither A nor B   (C5)

28. Technician A says that a nick on any of the crankshaft journals that can be caught with a fingernail has to be serviced. Technician B says that crankshafts can be checked for cracks magnetically by using magnetic particle inspection. Who is right?
    A. A only
    B. B only
    C. Both A and B
    D. Neither A nor B   (C6)

29. When inspecting the old crankshaft bearings, all of the following can be determined EXCEPT:
    A. mileage.
    B. crankshaft misalignment.
    C. lack of lubrication.
    D. metal-to-metal contact in the engine.   (C7)

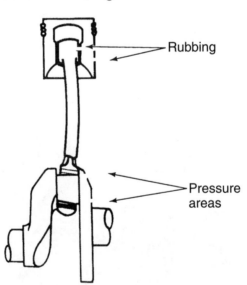

30. Referring to the figure above, a bent connecting rod may cause all of these EXCEPT:
    A. uneven connecting rod bearing wear.
    B. uneven main bearing wear.
    C. uneven piston pin wear.
    D. uneven cylinder wall wear.   (C8)

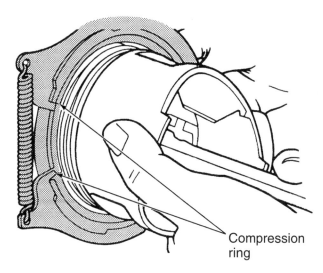

Compression ring

31. The tool in the figure above is being used for what purpose?
    A. To widen the piston ring grooves
    B. To deepen the piston ring grooves
    C. To remove and replace the piston rings
    D. To remove carbon from the piston ring grooves (C9)

32. The main reason for seized or scored piston pins is most likely which of these:
    A. a lack of lubrication.
    B. overheating of the piston.
    C. preignition.
    D. overspeeding the engine. (C10)

33. Technician A says that if the piston ring groove gap has exceeded specifications, that a larger piston ring must be used. Technician B says excessive piston ring clearance can cause a piston ring to break. Who is right?
    A. A only
    B. B only
    C. Both A and B
    D. Neither A nor B (C11)

34. When measuring the piston ring gap, as shown in the figure above, Technician A says that the ring gap should be measured with the ring positioned at the top of the ring travel in the cylinder. Technician B says the two compression rings are interchangeable on most pistons. Who is right?
    A. A only
    B. B only
    C. Both A and B
    D. Neither A nor B (C11)

35. Technician A says the vibration damper counterbalances the back-and-forth twisting motion of the crankshaft each time a cylinder fires. Technician B says if the seal contact area on the vibration damper hub is scored, the damper assembly must be replaced. Who is right?
    A. A only
    B. B only
    C. Both A and B
    D. Neither A nor B (C12)

36. All of the following are causes of low engine oil pressure **EXCEPT**:
    A. worn camshaft bearings.
    B. worn crankshaft bearings.
    C. weak oil pressure regulator spring tension.
    D. restricted pushrod oil passages. (D1)

37. The LEAST likely cause of oil pump failure is:
    A. oil pump gear lockup.
    B. oil pump drive rod failure.
    C. excessive engine speed.
    D. a defective oil pressure relief valve. (D2)

38. A loose alternator belt may cause:
    A. a discharged battery.
    B. a squealing noise while decelerating.
    C. a damaged alternator bearing.
    D. engine overheating. (D3)

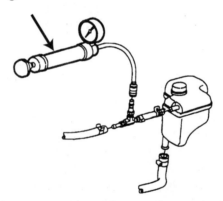

39. The tester in the figure above may be used to test the following items **EXCEPT**:
    A. cooling system leaks.
    B. radiator cap pressure relief valve.
    C. coolant specific gravity.
    D. heater core leaks. (D4)

40. A vehicle has an overheating problem. The vehicle sometimes overheats within the first five minutes after the engine is started. Other times, the cooling system operates normally. Technician A says that the water pump can be working intermittently. Technician B says that the thermostat is sometimes sticking and needs replacement. Who is right?
    A. A only
    B. B only
    C. Both A and B
    D. Neither A nor B (D5)

41. While discussing cooling system flushing, Technician A says that to properly flush the radiator, reverse flushing should be performed to dislodge deposits. Technician B says that a flush involves draining the coolant and refilling the system with new antifreeze. Who is right?
    A. A only
    B. B only
    C. Both A and B
    D. Neither A nor B (D6)

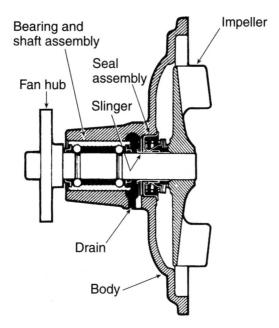

42. Referring to the figure above, what is the most common reason for replacing the water pump?
    A. The pump impeller becomes deteriorated from corrosion.
    B. The water pump to engine block gasket fails.
    C. The hub separates from the water pump shaft.
    D. Failure of the lip seal. (D7)

43. While removing a radiator from a vehicle, Technician A says that removing the lower hose first might make a mess, but the radiator drain petcock will not have to be opened. Technician B says that the first step to removing any radiator is to disconnect the negative battery cable. Who is right?
    A. A only
    B. B only
    C. Both A and B
    D. Neither A nor B (D8)

44. Technician A says that when the fluid is leaking out of the radiator fan clutch, the clutch must be replaced. Technician B says that all types of fan blades should be inspected for stress fractures. Who is right?
    A. A only
    B. B only
    C. Both A and B
    D. Neither A nor B (D9)

45. When diagnosing a no-start condition on a vehicle with no spark, Technician A says that the ignition module has malfunctioned and should be replaced. Technician B says that the hall effect switch in the distributor has lost its polarity and needs replacement. Who is right?
    A. A only
    B. B only
    C. Both A and B
    D. Neither A nor B (E1)

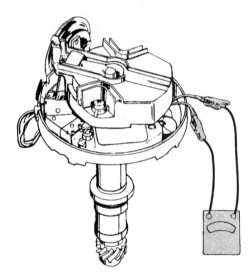

46. In the figure above, the ohmmeter reads an infinite reading in the auto range setting. The pickup coil has which of these conditions?
    A. Short to ground
    B. Shorted
    C. Grounded
    D. Open (E2 and E3)

47. An ignition system that is sensitive to moisture could be caused by all of the following **EXCEPT:**
    A. a cracked distributor cap.
    B. deteriorated ignition wires.
    C. a broken ignition coil core.
    D. loose distributor mounting. (E3 and E4)

48. Technician A says that when an ignition coil fails, it will leak oil from the tower. Technician B says that ignition coils are part of the primary and secondary ignition circuits. Who is right?
    A. A only
    B. B only
    C. Both A and B
    D. Neither A nor B (E4)

49. If certain spark plugs fire and others do not and the secondary circuit is suspected to be the cause, which of the following could be the cause?
    A. Ignition switch
    B. Hall effect switch
    C. Ignition coil
    D. Spark plug wires (E4)

50. What type of tool is most commonly used when checking an ignition coil?
    A. OEM specific scan tool
    B. Digital multimeter (DMM)
    C. An oscilloscope
    D. Test light (E5)

51. When adjusting the ignition timing on a vehicle, Technician A says that the underhood emission label has specific instructions for setting the timing. Technician B says that on non-feedback style carburetors, the engine idle speed must be verified before the timing can be adjusted. Who is right?
    A. A only
    B. B only
    C. Both A and B
    D. Neither A nor B (E6)

52. While discussing ignition point adjustment, Technician A says an increase in point gap increases the cam dwell reading. Technician B says point dwell must be long enough to allow the magnetic field to build up in the coil. Who is right?
    A. A only
    B. B only
    C. Both A and B
    D. Neither A nor B (E7)

53. What does an ignition module tester analyze?
    A. It tests the secondary ignition circuit.
    B. It tests the primary ignition circuit.
    C. It tests the available voltage to the ignition module.
    D. It checks the ability of the module to turn on and off. (E8)

54. While diagnosing a vehicle with a no-start condition that has been narrowed down to the fuel system, Technician A says that the fuel level should be checked first. Technician B says the injector pulse width should be checked. Who is right?
    A. A only
    B. B only
    C. Both A and B
    D. Neither A nor B (F1)

55. A vehicle that is experiencing poor fuel economy has entered the shop. Technician A says that if there is no fuel leaking from the fuel pump, then there is no problem with the pump. Technician B says that the fuel pump could be leaking internally and the engine oil level should be checked to help indicate a problem. Who is right?
    A. A only
    B. B only
    C. Both A and B
    D. Neither A nor B (F2)

56. What is the last component to be disconnected when removing the carburetor?
    A. The throttle linkage
    B. Transmission kickdown lever
    C. The fuel supply line
    D. The brake booster vacuum supply hose (F3)

57. Technician A says that the solution used to clean carburetors is corrosive and the plastic components should be removed beforehand. Technician B says that feedback style carburetors have special coatings on the throttle plates and they cannot be cleaned with solvents. Who is right?
    A. A only
    B. B only
    C. Both A and B
    D. Neither A nor B (F4)

58. When replacing a throttle body on a vehicle, Technician A says the PCV valve supply hose is connected to the throttle body. Technician B says that the throttle body gasket can be reused. Who is right?
    A. A only
    B. B only
    C. Both A and B
    D. Neither A nor B   (F5)

59. Technician A says that intake manifold gaskets can be replaced with silicone. Technician B says that intake manifold gaskets can be bought in a kit with all other gaskets in the intake system. Who is right?
    A. A only
    B. B only
    C. Both A and B
    D. Neither A nor B   (F6)

60. What is the function of the automatic choke?
    A. To limit the amount of air entering the engine
    B. To provide blowby vapor for the PCV system
    C. To provide the best possible engine performance in cold weather
    D. To add raw fuel to the engine during cold startup   (F7)

61. Technician A says that closed loop operation is when the engine is at operating temperature, and the PCM uses the information from the oxygen sensor to control the air-fuel ratio. Technician B says that injector dwell is rated in seconds. Who is right?
    A. A only
    B. B only
    C. Both A and B
    D. Neither A and B   (F8)

62. During deceleration, what is the first sensor value to change?
    A. Manifold absolute pressure (MAP) sensor
    B. Transmission output shaft speed sensor
    C. Mass airflow (MAF) sensor
    D. Throttle position sensor (TPS)   (F9)

63. Technician A says that air cleaner elements on newer fuel-injected vehicles can last for 50,000 to 75,000 miles before needing replacement. Technician B says that a shop light can be used to determine how dirty the element is. Who is right?
    A. A only
    B. B only
    C. Both A and B
    D. Neither A nor B   (F10)

64. Technician A says that the fuel pressure regulator senses engine vacuum. Technician B says the fuel pressure regulator controls the amount of fuel that is returned to the tank. Who is right?
   A. A only
   B. B only
   C. Both A and B
   D. Neither A nor B (F11)

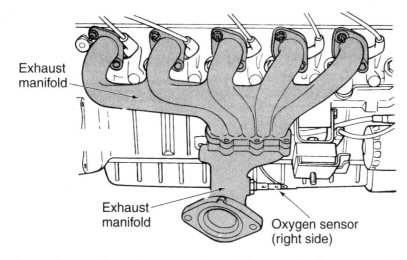

65. In the figure above, what advantages do stainless steel exhaust manifolds have over conventional cast iron manifolds?
   A. Lower in cost
   B. Heavier in weight
   C. Higher emissions
   D. Lighter in weight (F12)

66. Technician A says that a battery can be charged at any rate as long as the electrolyte does not boil and the temperature does not exceed 120°F (50°C). Technician B says fast charging is just as complete as charging a battery at a slower rate. Who is right?
   A. A only
   B. B only
   C. Both A and B
   D. Neither A nor B (G2)

67. All these statements about battery cable service are true **EXCEPT:**
   A. the terminals should be removed from the battery with a puller.
   B. remove the negative battery cable first.
   C. remove the positive battery cable before the negative cable.
   D. cleanse the battery with a baking soda and water solution. (G5)

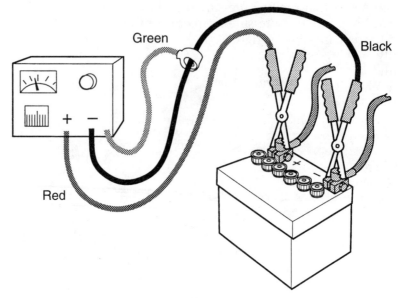

68. In the figure above, what is being tested?
    A. Alternator current draw test
    B. Starter current draw test
    C. Battery load test
    D. Ground circuit connection (G3)

69. Technician A says that when jump starting a vehicle with a dead battery, place rags over the battery to be boosted. Technician B says to connect the negative cables first to prevent sparks. Who is right?
    A. A only
    B. B only
    C. Both A and B
    D. Neither A nor B (G4)

70. Technician A says that battery terminals can be treated with petroleum jelly to help prevent corrosion. Technician B says that protective pads can be placed on the terminals to prevent corrosion. Who is right?
    A. A only
    B. B only
    C. Both A and B
    D. Neither A nor B (G5)

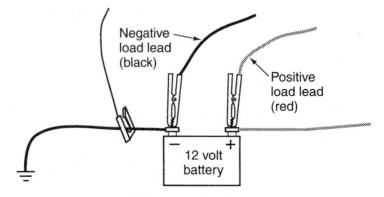

71. In the figure above, what procedure is being performed?
    A. Starter solenoid current draw
    B. Starter motor current draw
    C. Battery load test
    D. Resistance check (G6)

72. Technician A says that a starting and charging system tester can be used to test free spinning current draw tests. Technician B says that free spinning current draw tests must be performed with the starter solenoid removed. Who is right?
    A. A only
    B. B only
    C. Both A and B
    D. Neither A nor B (G7)

73. All of the following about EGR valves is true **EXCEPT:**
    A. they operate on ported venturi vacuum systems.
    B. they operate on ported vacuum.
    C. they may be mounted on the intake.
    D. they operate at wide open throttle. (H1)

74. Technician A says that the coolant temperature override switch limits coolant flow to the EGR valve until the engine warms up to operating temperature. Technician B says the EVR solenoid converts the electrical signals from the PCM into a mechanical action which directs vacuum to the EGR valve. Who is right?
    A. A only
    B. B only
    C. Both A and B
    D. Neither A nor B (H2)

75. When troubleshooting the air-injection system, the first thing to be inspected should be the:
    A. air pump drive belt.
    B. air-injection hoses.
    C. the hose from the pump.
    D. three-way catalytic converter. (H3)

76. A vehicle with a seized air-injection pump has entered the shop for service. Technician A says that the air pump needs to be rebuilt. Technician B says that you should never pry on the housing while tightening the drive belt. Who is right?
    A. A only
    B. B only
    C. Both A and B
    D. Neither A nor B (H4)

77. Technician A says that the gulp valve is used to prevent engine backfires. Technician B says the gulp valve has largely been replaced by the diverter valve? Who is correct?
    A. A only
    B. B only
    C. Both A and B
    D. Neither A nor B (H5)

78. What type of hose can be used to replace air-injection hoses?
    A. Preformed air-injection hose
    B. Preformed heater hose
    C. Fuel supply hose
    D. All of the above (H6)

79. Which of the following are indications that the PCV system is not functioning properly?
    A. No oil consumption
    B. Normal HC and CO emission readings
    C. Normal fuel consumption
    D. Excessive blowby (H7)

80. Technician A says that all three-way catalytic (TWC) converters are interchangeable. Technician B says that an overly rich air-fuel mixture will not affect the catalytic converter. Who is right?
    A. A only
    B. B only
    C. Both A and B
    D. Neither A nor B (H8)

81. A fuel-injected engine has a severe surging problem only at 55 mph (88 km/h) or faster. Engine operation is normal at idle and low speeds. Technician A says there may be low voltage at the fuel pump. Technician B says that the inertia switch may have high resistance. Who is right?
    A. A only
    B. B only
    C. Both A and B
    D. Neither A nor B (I1)

82. When diagnosing an EVAP system with a scan tool, the PCM never provides the on command to the EVAP solenoid at any engine or vehicle speed. Technician A says to check the ECT sensor signal to the PCM. Technician B says check the vacuum hoses from the intake to the EVAP canister. Who is right?
    A. A only
    B. B only
    C. Both A and B
    D. Neither A nor B (I2)

83. While discussing ECT sensor diagnosis, Technician A says a defective ECT sensor may cause hard cold starting. Technician B says a defective ECT sensor may cause improper emissions. Who is right?
    A. A only
    B. B only
    C. Both A and B
    D. Neither A nor B (I3)

84. While using a DMM to diagnose a throttle position sensor, Technician A says that the sensor can be checked with the meter on the resistance scale or the voltage scale. Technician B says that if the PCM does not set a throttle position sensor fault code, there is no problem with the sensor. Who is right?
    A. A only
    B. B only
    C. Both A and B
    D. Neither A nor B (I4)

85. Technician A says that most manufacturers provide technical service information on compact discs (CD). Technician B says that if available, service manuals from the vehicle's manufacturer have more detailed information regarding procedures and diagnostic information. Who is right?
    A. A only
    B. B only
    C. Both A and B
    D. Neither A nor B (I5)

86. Technician A says that circuit protection devices open when the current in the circuit falls below a predetermined level. Technician B says excessive current results from a decrease in the circuit's resistance. Who is right?
    A. A only
    B. B only
    C. Both A and B
    D. Neither A nor B (I6)

6 Additional Test Questions for Practice

Additional Test Questions

Please note the letter and number in parentheses following each sample question. They match the overview in section 4 that discusses the relevant subject matter. You may want to refer to the overview using this cross-referencing key to help with questions posing problems for you.

1. What should be done to a customer's vehicle after it has been repaired?
   A. Take the vehicle on a road test.
   B. Top off all the fluids.
   C. Provide paper floor mats.
   D. Underhood detailing. (A1)

2. A cooling system is pressurized with a pressure tester to locate a coolant leak. After 15 minutes, the tester gauge has dropped from 15 to 5 psi, and there are no visible signs of coolant leaks in the engine compartment. Technician A says the engine may have a leaking head gasket. Technician B says that the heater core may be leaking. Who is right?
   A. A only
   B. B only
   C. Both A and B
   D. Neither A nor B (A2)

3. A heavy thumping machine gun sounding type noise occurs with the engine idling, but the oil pressure is normal. The noise may be caused by:
   A. worn pistons.
   B. loose flywheel bolts.
   C. worn main bearings.
   D. loose camshaft bearings. (A4)

4. What two things can engine coolant be tested for?
   A. Carbon and sulfur content
   B. Lead content and viscosity
   C. Acidity and freezing point
   D. Age and water pump condition (A3)

5. A fuel-injected vehicle has a steady puff noise coming from the exhaust with the engine idling. The cause could be:
   A. a burned exhaust valve.
   B. excessive fuel pressure.
   C. a restricted fuel return line.
   D. a sticking fuel pump check valve. (A5)

6. With a vacuum gauge connected to manifold vacuum, the needle fluctuates between 15 and 20 in. Hg. What could be the cause?
   A. Late ignition timing
   B. Intake manifold gasket failure
   C. A restricted exhaust system
   D. Sticking valve stems and guides (A6)

7. To perform a cylinder compression test, the first step is:
   A. charge the battery.
   B. warm up the engine to operating temperature.
   C. disable the ignition and the fuel system.
   D. hold the accelerator to the floor. (A8)

8. During a cylinder leakage test, cylinder number 4 has 50 percent leakage and the air can be heard escaping from the PCV valve opening. Technician A says that the intake valve in that cylinder is leaking. Technician B says that the piston rings in that cylinder are worn. Who is right?
   A. A only
   B. B only
   C. Both A and B
   D. Neither A nor B (A9)

9. Technician A says that a fouled spark plug will cause a higher firing voltage. Technician B says that circuit gap or firing line voltage is the voltage required to fire all the spark plugs in the secondary circuit. Who is right?
   A. A only
   B. B only
   C. Both A and B
   D. Neither A nor B (A10)

10. During a cylinder leakage test, Technician A says that the cylinder is filled with regulated shop air. Technician B says that both intake and exhaust valves must be closed. Who is right?
    A. A only
    B. B only
    C. Both A and B
    D. Neither A nor B (A9)

11. Valve seat inspection is being discussed. Technician A says if the valve seat is not within specification, the cylinder head must be replaced. Technician B says seat run out is a measure of how circular the valve seat is in relation to the valve guide. Who is right?
    A. A only
    B. B only
    C. Both A and B
    D. Neither A nor B (B2)

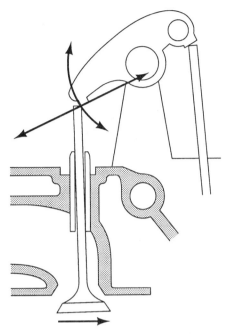

12. In the figure above, Technician A says the three high stress areas of the rocker arm are the pivot, pushrod contact, and valve stem contact surfaces. Technician B says if valve stem wear is not centered in the rocker arm, it may be caused by a bent stud. Who is right?
    A. A only
    B. B only
    C. Both A and B
    D. Neither A nor B   (B3)

13. Technician A says the valve stem must be measured at the top, middle, and near the fillet. Technician B says many modern engines use tapered stems to provide additional clearance between the stem and guide close to the head of the valve. Who is right?
    A. A only
    B. B only
    C. Both A and B
    D. Neither A nor B   (B4)

14. Reconditioning valves is being discussed. Technician A says the valve is stroked across the stone as the valve is being fed in. Technician B says when the last pass is complete you back the valve away from the stone. Who is right?
    A. A only
    B. B only
    C. Both A and B
    D. Neither A nor B   (B5)

15. Technician A says the amount removed from the valve face is the size of the required valve spring shim. Technician B says valve stem tip height must be correct for proper rocker arm geometry. Who is right?
    A. A only
    B. B only
    C. Both A and B
    D. Neither A nor B   (B5)

16. Technician A says after the valve springs are installed, attempt to turn the valve spring by hand. Technician B says if the valve springs are properly installed, the valve spring should rotate. Who is right?
    A. A only
    B. B only
    C. Both A and B
    D. Neither A nor B
    (B5)

17. Technician A says that a worn out valve guide will provide an inaccurate valve seat run out measurement. Technician B says that the valve seat to valve face contact area provides a path for heat from the valve head to dissipate. Who is right?
    A. A only
    B. B only
    C. Both A and B
    D. Neither A nor B
    (B7)

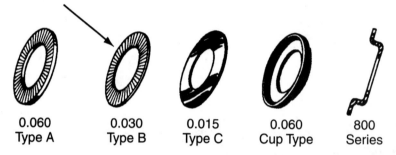

18. Technician A says if a valve spring shim is used, install the serrated side toward the spring. Technician B says to determine required shim thickness, the measured distance is subtracted from the specifications. Who is right?
    A. A only
    B. B only
    C. Both A and B
    D. Neither A nor B
    (B8)

19. Technician A says that the pushrods can be checked for straightness by rolling them on a known flat surface. Technician B says that if a pushrod is bent, but it cannot be seen by the eye, it will not cause a problem. Who is right?
    A. A only
    B. B only
    C. Both A and B
    D. Neither A nor B
    (B9)

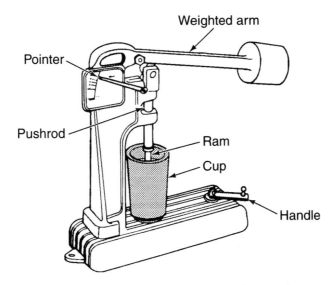

20. In the figure above, Technician A says that the tool is used to prime the lifters. Technician B says that the tool is used to determine lifter leak down. Who is right?
    A. A only
    B. B only
    C. Both A and B
    D. Neither A nor B   (B10)

21. When using room temperature vulcanizing (RTV) for gasket applications, which of the following is most true?
    A. The gasket surfaces should be cleaned with carburetor cleaner.
    B. The RTV should be placed on one side of the bolt holes only.
    C. The RTV should be allowed to cure for 25 minutes before assembly.
    D. The RTV bead should be 1/8 in. (3 mm) in width.   (C1)

22. All of the following are reasons for engine block fractures **EXCEPT:**
    A. fatigue.
    B. loss of oil pressure.
    C. impact damage.
    D. detonation.   (C2)

23. While discussing cylinder taper, Technician A says taper in the bore causes the ring end gaps to change while the piston moves in the bore. Technician B says a cylinder bore gauge can be used to determine taper. Who is right?
    A. A only
    B. B only
    C. Both A and B
    D. Neither A nor B   (C3)

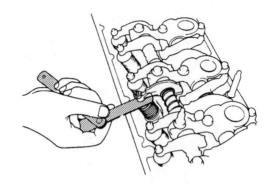

24. In the figure above, Technician A says that feeler gauge should slide with slight resistance. Technician B says that on some applications, the intake valve clearance is less than the clearance for the exhaust valve. Who is right?
   A. A only
   B. B only
   C. Both A and B
   D. Neither A nor B (B11)

25. While removing the cylinder wall ridge, which of the following should be performed?
   A. Place an oiled rag in the cylinder to catch any metal shavings.
   B. After removing the ridge, smooth the surface with 80 grit sandpaper.
   C. If the ridge has a step less than 0.050 in. (1.2 mm), there is no need for it to be removed.
   D. Wash the cylinder walls with soap and water solution before removing the ridge. (C4)

26. When measuring camshaft bearing oil clearance on a overhead camshaft engine, Technician A says that "Plastigage®" can be placed between the camshaft journal and the bearing cap. Technician B says overhead camshaft bearings are made of aluminum and are inserts. Who is right?
   A. A only
   B. B only
   C. Both A and B
   D. Neither A nor B (C5)

27. Technician A says metal burrs on the crankshaft flange may cause excessive wear on the flexplate gear teeth. Technician B says metal burrs may cause improper torque converter to transmission alignment. Who is right?
   A. A only
   B. B only
   C. Both A and B
   D. Neither A nor B (C6)

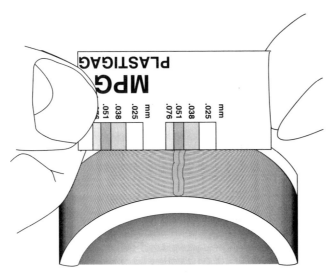

28. In the figure above, what is being performed?
    A. Removing a scratch in the bearing with crocus cloth
    B. Using a special tool to remove the bearing insert
    C. Measuring the thickness of the crushed Plastigage®
    D. Determining if the crankshaft journal has been machined   (C7)

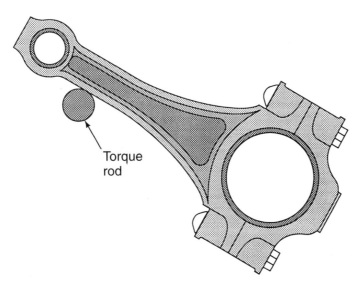

29. While observing the tool shown above, Technician A says that it is used to straighten the connecting rod bend. Technician B says that it is used to measure connecting rod twist and for connecting rod bend. Who is right?
    A. A only
    B. B only
    C. Both A and B
    D. Neither A nor B   (C8)

30. Removing press-fit piston pins is being discussed. Technician A says the pin can be driven out with a punch and a hammer while securing the piston in a vise. Technician B says removing the pin requires a press and special adapters. Who is right?
    A. A only
    B. B only
    C. Both A and B
    D. Neither A nor B   (C9)

31. Technician A says pistons are cam ground. Technician B says that pistons should be measured across the thrust surface of the skirt centerline of the piston pin. Who is right?
    A. A only
    B. B only
    C. Both A and B
    D. Neither A nor B (C10)

32. Technician A says when measuring a piston ring groove for wear, if the groove opening is larger than specifications, a thicker piston ring should be installed. Technician B says that excessive piston ring side clearance is not recommended, but it will not harm the piston or the rings. Who is right?
    A. A only
    B. B only
    C. Both A and B
    D. Neither A nor B (C11)

33. If a harmonic balancer outer ring has slipped on the rubber mounting, what must be done to correct it?
    A. Twist the ring back to its original position.
    B. Secure the outer ring by welding it in place.
    C. Replace the harmonic balancer.
    D. Press on a new outer ring. (C12)

34. The following are normal oil pump component measurements **EXCEPT:**
    A. inner rotor diameter.
    B. clearance between the rotors.
    C. inner and outer rotor thickness.
    D. outer rotor to housing clearance. (D1)

35. Technician A says that it is more cost effective to rebuild an oil pump than to buy a new one. Technician B says that it is a good practice to replace an oil pump when rebuilding an engine. Who is right?
    A. A only
    B. B only
    C. Both A and B
    D. Neither A or B (D2)

36. All of these are methods of measuring engine drive V-belt tension **EXCEPT:**
    A. use a belt tension gauge.
    B. measure the amount of belt deflection.
    C. visually see if the belt is contacting the bottom of the pulley.
    D. measure the length of the belt compared to a new one. (D3)

37. While discussing cooling systems, Technician A says that radiator pressure caps can be tested with a special adapter. Technician B says that pressure testing the cooling system can determine the condition of the thermostat. Who is right?
    A. A only
    B. B only
    C. Both A and B
    D. Neither A nor B (D4)

38. When filling a cooling system, Technician A says to fill half the system capacity with pure antifreeze and the remaining amount with water. Technician B says to use a 50/50 mixture of water to coolant mixture to fill the cooling system. Who is right?
    A. A only
    B. B only
    C. Both A and B
    D. Neither A nor B (D6)

39. Technician A says a defective water pump bearing may cause a growling noise when the engine is idling. Technician B says the water pump bearing may be ruined by coolant leaking past the pump seal. Who is right?
    A. A only
    B. B only
    C. Both A and B
    D. Neither A nor B (D7)

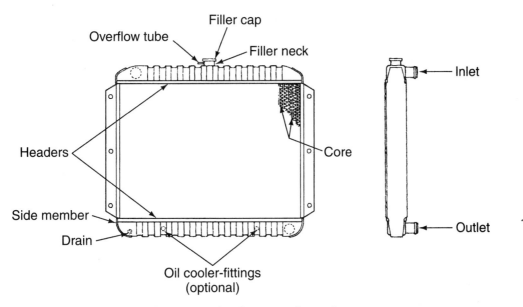

40. In the figure above, what type of radiator is shown?
    A. Downdraft
    B. Updraft
    C. Crossflow
    D. Downflow (D8)

41. The vacuum valve in the radiator cap is stuck closed. The result could be:
    A. collapsed upper radiator hose after the engine is shut off.
    B. excessive cooling system pressure at normal engine temperature.
    C. engine overheating when operating under heavy load.
    D. engine overheating during extended idle periods. (D9)

42. Which of the following would LEAST likely cause an engine misfire?
    A. Intake manifold leak
    B. Exhaust manifold leak
    C. Defective spark plugs
    D. Defective coil (E1)

43. Technician A says that with a 12-volt test lamp connected to the tach terminal, if the light flutters, the primary ignition system is functioning properly. Technician B says that the pickup coil cannot be tested with an ohmmeter. Who is right?
    A. A only
    B. B only
    C. Both A and B
    D. Neither A nor B (E2)

44. What should the mechanical advance unit be inspected for?
    A. To ensure that it moves when vacuum is applied
    B. Cracks in the diaphragm
    C. Brittle vacuum supply hose
    D. Broken or missing counterweight springs (E3)

45. Technician A says the secondary ignition circuit starts at the ignition switch and follows through to the spark plug. Technician B says the primary circuit is from the coil terminal through the distributor cap to the spark plugs. Who is right?
    A. A only
    B. B only
    C. Both A and B
    D. Neither A nor B (E4)

46. While testing an ignition coil with an ohmmeter, Technician A says the primary windings should have a resistance of about 0.5 to 2.0 ohms. Technician B says that the secondary circuit should have an infinite resistance. Who is right?
    A. A only
    B. B only
    C. Both A and B
    D. Neither A nor B (E5)

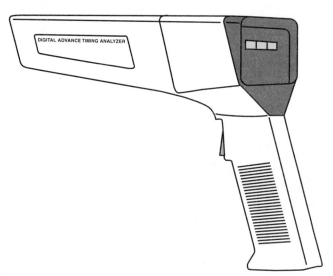

47. The timing light like the one in the figure above has what capability to do which of the following:
    A. be used as a trouble light.
    B. check spark advance.
    C. be used as an oscilloscope on the secondary ignition circuit.
    D. check the primary circuit ignition timing. (E6)

48. Technician A says that a short in the wiring inside the distributor could cause an open circuit in the pickup coil. Technician B says that most pickup coils have an infinite ohm reading. Who is right?
    A. A only
    B. B only
    C. Both A and B
    D. Neither A nor B (E7)

49. Technician A says that ignition module testers are universal and can be used with any module. Technician B says that each vehicle manufacturer has their own ignition module tester. Who is right?
    A. A only
    B. B only
    C. Both A and B
    D. Neither A nor B (E8)

50. While diagnosing a fuel related no-start condition, Technician A says that the fuel injectors are supplied with 12-volt power at all times. Technician B says that a fuel no-start is usually an electrical malfunction. Who is right?
    A. A only
    B. B only
    C. Both A and B
    D. Neither A nor B (F1)

51. Technician A says that electric fuel pumps are located in the tank for packing requirements. Technician B says that the fuel pumps run for 5 to 10 seconds during the ignition on cycle to pressurize the fuel rail during startup. Who is right?
    A. A only
    B. B only
    C. Both A and B
    D. Neither A nor B (F2)

52. If the technician is not familiar with removing a carburetor, which of the following should be done first?
    A. Disconnect the fuel supply line and start the engine to empty the float bowl.
    B. Use heat to help loosen any bolts that are rusted in place.
    C. Label all the connections so that reassembly will not be confusing.
    D. Ask another technician for advice. (F3)

53. While disassembling a carburetor, Technician A says this is a good time to check for signs of obvious damage. Technician B says that the subassemblies should be inspected at the same time. Who is right?
    A. A only
    B. B only
    C. Both A and B
    D. Neither A nor B (F4)

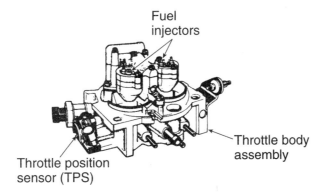

54. For what reason do some throttle bodies have a special coating?
    A. To prevent the engine from backfiring during deceleration
    B. So that the airflow is evenly distributed to all the cylinders evenly
    C. To prevent carbon buildup on the throttle plate
    D. To increase the velocity of the incoming air (F5)

55. All of the following are true about intake manifold gaskets **EXCEPT:**
    A. it's a good practice to replace all the intake manifold gaskets at the same time.
    B. the replacement gasket should be made of the same material as the original.
    C. conveniently, all the gaskets in the intake system come together in a kit.
    D. cracked intake manifold gaskets can be repaired with silicone. (F6)

56. When discussing automatic choke adjustment, Technician A says the choke must be inspected for proper operation before an adjustment can be made. Technician B says that the choke linkage must be free from dirt and debris. Who is right?
    A. A only
    B. B only
    C. Both A and B
    D. Neither A nor B   (F7)

57. Technician A says the PCM strategy monitors various sensors in order of importance. Technician B says that if the engine coolant temperature drops below a specified temperature, the PCM strategy decreases the injector pulse width. Who is right?
    A. A only
    B. B only
    C. Both A and B
    D. Neither A nor B   (F8)

58. The PCM can determine the vehicle is in the deceleration mode by which of the following sensors?
    A. Anti-lock brake sensor
    B. Throttle position sensor
    C. Canister purge solenoid
    D. Camshaft position sensor   (F9)

59. Technician A says that the optimum air fuel ratio is 18 to 1. Technician B says that for every gallon of gasoline that the engine burns, the engine uses 9,000 gallons of air. Who is right?
    A. A only
    B. B only
    C. Both A and B
    D. Neither A nor B   (F10)

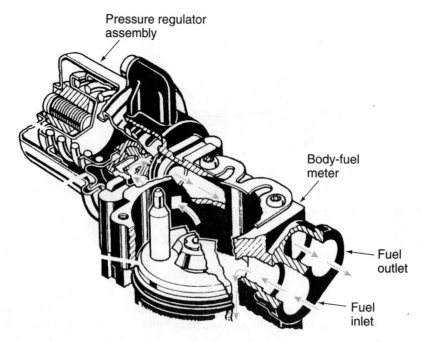

60. In the figure above, as engine manifold vacuum decreases, what does the spring in the fuel pressure regulator do?
    A. Open the valve to allow more fuel to the rail
    B. Open the valve to allow less fuel to return to the tank
    C. Close the valve to allow fuel pressure to build in the rail
    D. Close the valve so that the fuel pump will operate under a load   (F11)

61. While discussing exhaust manifolds, Technician A says that cast iron manifolds are significantly lighter than stainless steel. Technician B says that cast iron manifolds do not offer good resistance to change in temperature. Who is right?
    A. A only
    B. B only
    C. Both A and B
    D. Neither A nor B   (F12)

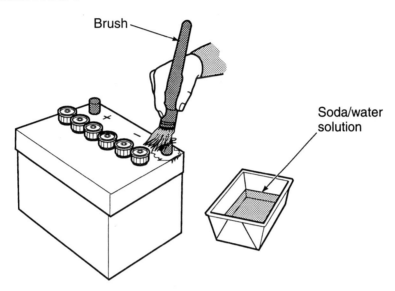

62. In the figure above, when discussing cleaning the battery, Technician A says the solution of baking soda and water should be allowed to run over the battery into the tray. Technician B says the leftover solution of water and baking soda can be used to fill the cells in the battery. Who is right?
    A. A only
    B. B only
    C. Both A and B
    D. Neither A nor B   (G1)

63. Technician A says that the normal temperature for a battery being charged is around 140°F (60°C). Technician B says the slower the battery is charged with a lower amp level, the more complete the charge. Who is right?
    A. A only
    B. B only
    C. Both A and B
    D. Neither A nor B   (G2)

64. When load testing a battery, how many amps should be placed on a 525 cold cranking amp battery?
    A. 125
    B. 325
    C. 262
    D. 425   (G3)

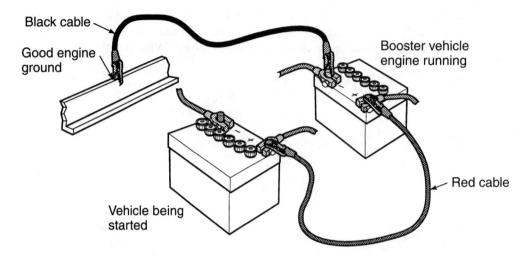

65. While jump starting a discharged battery, Technician A says to connect the negative battery cable first on the discharged battery. Technician B says that all the accessories must be turned off to prevent electrical damage. Who is right?
    A. A only
    B. B only
    C. Both A and B
    D. Neither A nor B (G4)

66. While discussing battery post service, Technician A says to use a wire brush to remove corrosion. Technician B says use the electrolyte solution to remove the oxidation from the posts. Who is right?
    A. A only
    B. B only
    C. Both A and B
    D. Neither A nor B (G5)

67. Technician A says that low cranking speed can be caused by high resistance in the starter motor circuit. Technician B says that internal engine problems can be the cause of the starter motor not functioning. Who is right?
    A. A only
    B. B only
    C. Both A and B
    D. Neither A nor B (G6)

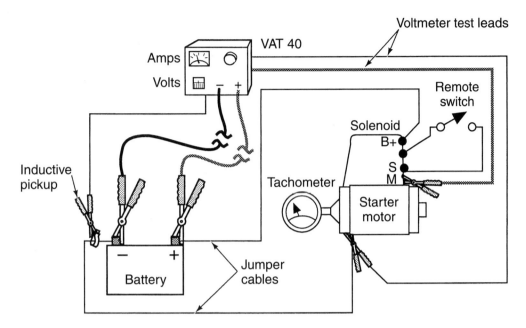

68. In the figure above, which of the following is being performed?
    A. Pinion gear clearance
    B. Checking the voltage drop across the ground circuit
    C. Battery load test
    D. Starter free speed test (G7)

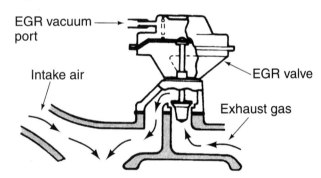

69. In the figure above, Technician A says that some EGR valves with heavy carbon buildup must be cleaned with a sandblaster. Technician B says to use fuel-injection cleaner to clean the passages in the EGR valve. Who is right?
    A. A only
    B. B only
    C. Both A and B
    D. Neither A nor B (H1)

70. While discussing EGR valves, Technician A says that the vacuum hoses usually do not cause failures with the EGR system. Technician B says that the plastic used for the EPT and EVR solenoids is very durable and will not crack around the vacuum hose connection. Who is right?
    A. A only
    B. B only
    C. Both A and B
    D. Neither A nor B (H2)

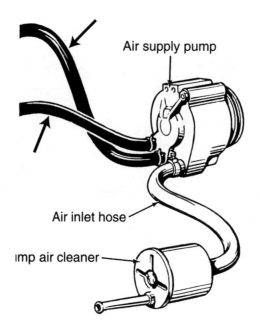

71. In the figure above, which of the following conditions would best describe what should happen when the two hoses indicated are disconnected while the engine is idling?
    A. A strong suction should be felt at the end of the disconnected hose.
    B. Either of the hoses should discharge airflow out the end of the hose.
    C. There should be no positive airflow under 4000 rpm.
    D. There should be no suction at the hoses until the air pump clutch is engaged.
    (H3)

72. When removing an air pump from a vehicle, Technician A says new air pumps will come with a new diverter valve. Technician B says that the new air pump may not come with a new pulley. Who is right?
    A. A only
    B. B only
    C. Both A and B
    D. Neither A nor B
    (H4)

73. After disconnecting the vacuum supply hose to the bypass valve, what should be present at the hose?
    A. No vacuum
    B. Positive air displacement
    C. Exhaust gas
    D. Vacuum
    (H5)

74. While replacing an air-injection manifold, which of the following should be done?
    A. Remove the exhaust manifold from the vehicle, then remove the air-injection manifold.
    B. Remove the air pump from the vehicle and then remove the air-injection manifold.
    C. Use high temperature silicone in place of the gasket.
    D. Apply penetrating oil to the nuts before removal.
    (H6)

75. Technician A says that regular oil change intervals prevent clogs in the PCV system. Technician B says to check the air filter for oil deposits and oil puddling. Who is right?
    A. A only
    B. B only
    C. Both A and B
    D. Neither A nor B
    (H7)

Additional Test Questions for Practice

76. Technician A says that the catalytic converter must be removed while it is still at operating temperature. Technician B says that the heat shields should be reinstalled after the catalytic converter is replaced. Who is right?
    A. A only
    B. B only
    C. Both A and B
    D. Neither A nor B   (H8)

77. Which of the following can occur if the oxygen sensor fails?
    A. The vehicle's cruise control will not engage
    B. The engine air-fuel ratio will be operating in a rich condition.
    C. The system will never go to closed loop.
    D. Excessive oxides of nitrogen will be produced.   (I1)

78. All of the following are true regarding the use of a scan tool **EXCEPT:**
    A. warm the engine to normal operating temperature.
    B. connect the power adapter to the cigar lighter.
    C. disconnect the negative battery cable before connecting the DLC connector.
    D. enter the model year, engine size, and vehicle model.   (I2)

79. All of the following are approved methods for diagnosing a ECT sensor **EXCEPT:**
    A. scan tool.
    B. digital multimeter (DMM).
    C. observing resistance values at a specified temperature.
    D. substituting a variable resistor.   (I3)

80. Digital multimeters can be used on all of the following scales **EXCEPT:**
    A. kiloamps.
    B. DC volts.
    C. AC volts.
    D. ohms.   (I4)

81. Technician A says that service manuals cannot anticipate all types of situations that may occur to a technician performing work on a vehicle. Technician B says service manuals are published in different languages. Who is right?
    A. A only
    B. B only
    C. Both A and B
    D. Neither A nor B   (I5)

82. How are circuits designed that are susceptible to overloads on a routine basis?
    A. Maxi-fuses
    B. A fusible link
    C. A circuit breaker
    D. Fuse   (I6)

83. While discussing customer driveability concerns, Technician A says to think of possible causes of the indicated problem. Technician B says you check the fuel level first. Who is right?
    A. A only
    B. B only
    C. Both A and B
    D. Neither A nor B   (A1)

84. Technician A says oil leaks are caused by gasket material being blown out of its sealing position. Technician B says that oil leaks can be found by adding a dye to the engine oil that is visible under ultraviolet light. Who is right?
    A. A only
    B. B only
    C. Both A and B
    D. Neither A nor B   (A2)

85. What can be determined with the engine coolant by checking it with pH strips?
    A. Freezing point
    B. Head gasket failure
    C. Acidity
    D. Coolant age (A3)

86. Which of the following can cause valve train noise?
    A. A retarded timing belt adjustment
    B. A loose crankshaft dampener bolt
    C. A loose serpentine drive belt
    D. Lack of lubrication (A4)

87. Which of the following exhaust conditions are the most likely indication of an engine misfire?
    A. Blue colored smoke from the tailpipe
    B. Excessive rattling of the exhaust system
    C. Cold exhaust temperature
    D. Puffing or wheezing (A5)

88. What type of component failure can cause vacuum gauge fluctuations between 7 and 20 in. Hg?
    A. Weak valve springs
    B. Burned intake valves
    C. A leaking head gasket
    D. Burned exhaust valves (A6)

89. While performing a cylinder power balance test, Technician A says that when all the cylinders are contributing equally to the engine power, all the cylinders will provide the specified rpm drop. Technician B says that often the reason for a cylinder not contributing to the engine power can be traced to the fuel or the ignition system. Who is right?
    A. A only
    B. B only
    C. Both A and B
    D. Neither A nor B (A7)

90. While performing a compression test, a gradual buildup of compression occurs with each stroke. Technician A says this is an indication of a burned exhaust valve. Technician B says that it indicates a cracked cylinder head. Who is right?
    A. A only
    B. B only
    C. Both A and B
    D. Neither A nor B (A8)

91. If an engine has a cylinder(s) with zero compression, another indication of this condition would be:
    A. a burned exhaust valve.
    B. excessive exhaust noise.
    C. excessive blowby.
    D. backfires through the exhaust. (A8)

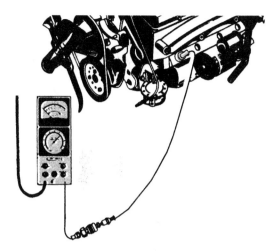

92. In the figure above, Technician A says that most leakage testers only have one air pressure gauge. Technician B says the second gauge indicates the air pressure in the cylinder. Who is right?
    A. A only
    B. B only
    C. Both A and B
    D. Neither A nor B   (A9)

93. How can it be determined during a cylinder leakage test if one or more of the intake valves are not sealing?
    A. Air can be heard escaping from the tailpipe.
    B. Air can be heard escaping from the PCV valve opening.
    C. Air escaping from the throttle body opening or from the carburetor
    D. Bubbles in the engine coolant   (A9)

94. Technician A says primary circuit test functions on an engine analyzer include coil input voltage, coil primary resistance, curb idle, and idle vacuum. Technician B says that the kV test measures the voltage required to jump the spark plug air gap. Who is right?
    A. A only
    B. B only
    C. Both A and B
    D. Neither A nor B   (A10)

95. What is the purpose of the snap test function on an engine analyzer?
    A. To check the voltage increase for each cylinder under load
    B. To check if all the spark plugs have the same air gap
    C. To indicate the condition of the primary circuit switching device
    D. To observe the ratio between engine speed and the firing kV   (A10)

96. A cylinder head that has been removed from a vehicle has carbon in the combustion chamber. Technician A says that this is not a normal condition and it should be looked into further. Technician B says that if the carbon is excessive it should be cleaned. Who is right?
    A. A only
    B. B only
    C. Both A and B
    D. Neither nor B   (B1)

97. Technician A says that no matter what condition the cylinder is in, it is always more cost effective to repair the cylinder head rather than replacing it. Technician B says that welding cylinder head cracks can save the customer the cost of a new or remanufactured one. Who is right?
    A. A only
    B. B only
    C. Both A and B
    D. Neither A nor B   (B1)

98. While discussing valve spring testing, Technician A says that spring tension should be within 30 percent of the manufacturer's specifications. Technician B says that there should be no more than 10 pounds difference between the springs. Who is right?
    A. A only
    B. B only
    C. Both A and B
    D. Neither A nor B   (B2)

99. What is most likely to happen if a valve spring is not square?
    A. The valve stem will warp.
    B. The valve spring will crack.
    C. The valve guide will be side loaded and cause premature wear.
    D. The free length of the valve spring will be altered.   (B2)

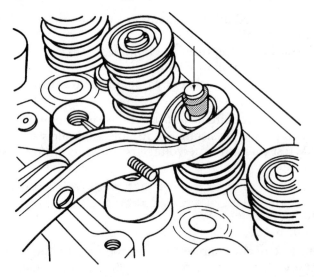

100. In the figure above, while inspecting the engine valves, what should the valve lock grooves be inspected for?
    A. Carbon deposits
    B. Thickness of grooves
    C. Flat spots on the valve face
    D. Uneven shoulders   (B3)

Additional Test Questions for Practice   Additional Test Questions   93

101. In the figure above, what is the purpose of the function being performed?
    A. To measure valve margin thickness
    B. To indicate valve guide wear
    C. To measure valve seat concentricity
    D. To measure valve seat run out   (B4)

102. While resurfacing the engine valves, what can indicate if the grinding stone needs to be dressed?
    A. When pieces of the dressing stone are missing
    B. After using the stone to condition 40 to 60 valves
    C. If the oil supply does not cool the stone properly
    D. If the valve chatters during the resurfacing process   (B5)

103. All of the following are true regarding valve seat resurfacing **EXCEPT:**
    A. lift the grinding stone on and off at the rate of 120 times per minute.
    B. worn valve guides must be replaced before attempting to grind any.
    C. only remove enough material to provide a new surface.
    D. apply pressure to the grinding stone.   (B6)

104. Technician A says to measure the valve seat width with a machinist's ruler. Technician B says to remove material according to the valve seat width measurement. Who is right?
    A. A only
    B. B only
    C. Both A and B
    D. Neither A nor B   (B7)

105. While assembling the valve train components, Technician A says to polish the valve stems with 240 grit sandpaper. Technician B says to lubricate the valve stems with assembly lubricant, or at least with clean engine oil. Who is right?
    A. A only
    B. B only
    C. Both A and B
    D. Neither A nor B   (B8)

106. Rocker arms can be made of all of the following **EXCEPT:**
    A. cast iron.
    B. stamped steel.
    C. titanium.
    D. aluminum.   (B9)

107. While inspecting a rocker arm shaft it is found to be galled and scored. Technician A says that wear in that location indicates there was a lack of lubrication. Technician B says that the shaft itself cannot be checked for straightness. Who is right?
   A. A only
   B. B only
   C. Both A and B
   D. Neither A nor B

(B9)

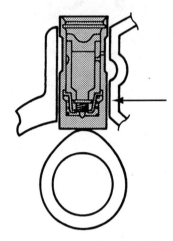

108. In the figure above, Technician A says that the lifter is a solid roller type. Technician B says that the base of the lifter should be inspected for the wear pattern. Who is right?
   A. A only
   B. B only
   C. Both A and B
   D. Neither A nor B

(B10)

109. What is the reason for varying the engine speed during the break-in period for flat tappet lifters?
   A. To ensure that the oil comes to operating temperature completely
   B. So that the thermostat opens quickly
   C. So enough oil is cooling and lubricating the contact surfaces of the lifter
   D. So the break-in period is over sooner

(B10)

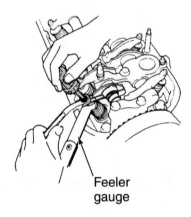

Feeler gauge

110. In the figure above, what is the arrangement used to adjust the valve clearance?
   A. Adjusting nut that retains the rocker arm to a rocker arm stud
   B. An adjustable screw that follows the camshaft
   C. Adjustable pushrods
   D. An adjustable screw in the end of the rocker arm

(B11)

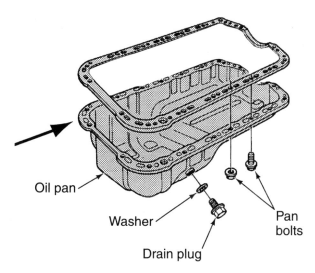

111. In the figure above, Technician A says stamped steel gasket surfaces can be straightened. Technician B says to grease the gasket mating surfaces on top to provide surface adhesion. Who is right?
    A. A only
    B. B only
    C. Both A and B
    D. Neither A nor B   (C1)

112. A powertrain control module (PCM) must be replaced in a vehicle. Technician A says a ground strap and conductive mat must be used to ensure that there is no static electricity damage to the vehicle. Technician B says disconnecting the vehicle's battery and touching bare metal on the vehicle prior to replacement will protect the vehicle from static electricity damage. Who is right?
    A. A only
    B. B only
    C. Both A and B
    D. Neither A or B   (I7)

113. When should a ground strap and conductive mat be used?
    A. When performing battery service
    B. When working on the secondary ignition system
    C. When replacing the ignition coil
    D. When working with electronic control modules   (I7)

Appendices

Answers to the Test Questions for the Sample Test Section 5

1.	C	23.	B	45.	C	67.	C
2.	A	24.	C	46.	D	68.	C
3.	B	25.	D	47.	D	69.	D
4.	D	26.	D	48.	B	70.	C
5.	C	27.	A	49.	D	71.	B
6.	B	28.	C	50.	B	72.	A
7.	C	29.	A	51.	C	73.	D
8.	D	30.	B	52.	B	74.	B
9.	D	31.	C	53.	D	75.	A
10.	B	32.	B	54.	C	76.	B
11.	A	33.	B	55.	B	77.	C
12.	D	34.	B	56.	C	78.	A
13.	C	35.	A	57.	A	79.	D
14.	B	36.	D	58.	D	80.	D
15.	A	37.	C	59.	B	81.	C
16.	C	38.	A	60.	C	82.	A
17.	C	39.	D	61.	A	83.	C
18.	A	40.	B	62.	D	84.	A
19.	D	41.	C	63.	B	85.	C
20.	C	42.	D	64.	C	86.	B
21.	C	43.	C	65.	D		
22.	C	44.	B	66.	A		

Explanations to the Answers for the Sample Test Section 5

Question #1
Answer A is a good choice because customer complaints must be listened to closely. Yet, it is wrong because both technicians are right.
Answer B is a good choice because when diagnosing a vehicle problem, always start with the easiest test to perform. Yet, it is wrong because both technicians are right.
Answer C is correct because both technicians are right.
Answer D is wrong because both technicians are right.

Question #2
Answer A is correct because the engine must be steam cleaned before an accurate diagnosis for an oil leak can be given.
Answer B is wrong because even if the lower portion of the engine is leaking, it can be areas on the top of the engine, rear main seal, or cam galley plugs as well as the oil pan gasket.
Answer C is wrong because only one technician is right.
Answer D is wrong because one technician is right.

Question #3
Answer A is wrong because there is no way to accurately indicate engine age without disassembly.
Answer B is correct because if the oil is dark and dirty, it has exceeded the change interval.
Answer C is wrong because there is no indication other than the service history.
Answer D is wrong because oil color does not indicate bearing wear.

Question #4
Answer A is wrong because bent pushrods do not make a knocking noise.
Answer B is wrong because collapsed lifters also will not make a knocking noise.
Answer C is wrong because a lack of oil pressure to the valvetrain would make tapping noises.
Answer D is correct because a worn main bearing would make a knocking noise when first started.

Question #5
Answer A is wrong because worn valve guides make blue smoke.
Answer B is wrong because worn valve seals also make blue smoke.
Answer C is correct because a fouled spark plug will not allow the fuel to burn resulting in black smoke, therefore, the exception.
Answer D is wrong because worn piston rings will make blue smoke.

Question #6
Answer A is wrong because if the engine has a blocked catalytic converter or exhaust, the reading would decrease not increase as stated.
Answer B is correct because if the vacuum increases as stated, the engine is functioning properly.
Answer C is wrong because only one technician is right.
Answer D is wrong because one technician is right.

Question #7
Answer A is a good choice because an intake manifold leak can cause a misfire at low rpm or idle and can disappear at higher engine speeds. Yet, it is wrong because both technicians are right.
Answer B is a good choice because engines sometimes stumble at idle and that retorquing the manifold will solve the problem. Yet, it is wrong because both technicians are right.
Answer C is correct because both technicians are right.
Answer D is wrong because both technicians are right.

Appendices Explanations to the Answers for the Sample Test Section 5   99

Question #8
Answer A is wrong because an oversized cylinder will have the same compression as the others.
Answer B is wrong because oil in the cylinder can only seal the piston rings.
Answer C is wrong because both technicians are wrong.
Answer D is correct because neither technician is right.

Question #9
Answer A is wrong because 5 to 10 percent is too low.
Answer B is wrong because 30 to 40 percent is too high.
Answer C is wrong because 40 to 45 percent is too high.
Answer D is correct because 15 to 20 percent is the correct tolerance.

Question #10
Answer A is wrong because engine analyzers and oscilloscopes are not obsolete and with adapters can be used on distributorless ignition.
Answer B is correct because scopes and engine analyzers are used to check out secondary circuits on all spark ignited engines.
Answer C is wrong because only one technician is right.
Answer D is wrong because one technician is right.

Question #11
Answer A is correct because the head needs to be resurfaced. When measuring the cylinder head for warpage with a straightedge, the feeler gauge measurement is 0.025 in. (0.63 mm).
Answer B is wrong because the measurement is out of tolerance. The standard is .003 inch for every 6 inches.
Answer C is wrong because only one technician is right.
Answer D is wrong because one technician is right.

Question #12
Answer A is wrong because springs should be inspected for cracks.
Answer B is wrong because valve springs' spring tension should be inspected.
Answer C is wrong because pitting and nicks are unwanted.
Answer D is correct because spring coil gap is the exception and not a standard valve spring measurement.

Question #13
Answer A is a good choice because you do inspect the valve spring retainers for cracks and discoloration. Yet, it is wrong because both technicians are right.
Answer B is a good choice because it is a good practice to replace the valve spring retainers when the valves are serviced. Yet, it is wrong because both technicians are right.
Answer C is correct because both technicians are right.
Answer D is wrong because both technicians are right.

Question #14
Answer A is wrong because the guides do not need to be lubricated.
Answer B is correct because you do clean the valve guide with a bore brush.
Answer C is wrong because many valve guides cannot be reamed.
Answer D is wrong because the valve guide does not need to be held in place.

Question #15
Answer A is correct because on occasion the valve stem requires resurfacing.
Answer B is wrong because many engines use an interference fit that comes from having a degree difference between the seat and valve faces.
Answer C is wrong because many valve guides cannot be reamed.
Answer D is wrong because the valve guide does not need to be held in place.

Question #16
Answer A is wrong because you do dress the stone before cutting any seat.
Answer B is also wrong because you do remove only enough material to provide a new surface.
Answer C is correct because you do not use transmission fluid to lubricate the stone, only special valve cutting oil.
Answer D is wrong because you do not apply pressure to the grinding stone.

Question #17
Answer A is a good choice because proper valve seat concentricity is critical for a proper seal, and a usual tolerance for seat run out is 0.002 in. (.50 mm). Yet, it is wrong because both technicians are right.
Answer B is a good choice. because the valve to seat contact is critical to an airtight seat and provides a path for the heat to dissipate. Yet, it is wrong because both technicians are right.
Answer C is correct because both technicians are right.
Answer D is wrong because both technicians are right.

Question #18
Answer A is correct because you correct valve stem height by removing material from the valve tip.
Answer B is wrong because shims are used for assembled spring height.
Answer C is wrong because this will not correct the length of the valve stem itself.
Answer D is wrong because this will not correct the length of the valve stem itself.

Question #19
Answer A is wrong because the pushrod contact receives excessive stress.
Answer B is wrong because the pivot is also under stress.
Answer C is wrong because the valve stem contact is under stress.
Answer D is correct because the main body is the base and receives the least stress.

Question #20
Answer A is wrong because a convex counter-machined face is not appropriate for a valve lifter.
Answer B is wrong because a smooth will not retain oil for lubrication.
Answer C is correct because a centered circular wear pattern is most desirable.
Answer D is wrong because a crosshatched pattern is mostly seen in cylinder walls and would be too coarse for a valve lifter.

Question #21
Answer A is a good choice because when the valve clearance is checked on a cylinder, that cylinder should be positioned at TDC on the exhaust stroke. Yet, it is wrong because both technicians are right.
Answer B is a good choice because some mechanical lifters have removable shim pads available in various thickness to provide proper valve clearances. Yet, it is wrong because both technicians are right.
Answer C is correct because both technicians are right.
Answer D is wrong because both technicians are right.

Question #22
Answer A is a good choice because the oil leak cause could be a crack or a rusted hole in the pan. Yet, it is wrong because both technicians are right.
Answer B is a good choice because the steel flange that mates to the engine block could cause the leak. Yet, it is wrong because both technicians are right.
Answer C is correct because both technicians are right.
Answer D is wrong because both technicians are right.

Question #23
Answer A is wrong because in most cases one can repair a stamped steel oil pan.
Answer B is correct because the flange can be straightened by striking the surface with a ball peen hammer on a true flat surface.
Answer C is wrong because only one technician is right.
Answer D is wrong because one technician is right.

Question #24
Answer A is a good choice because a warped cylinder head mounting surface on an engine block may cause valve seat distortion. Yet, it is wrong because both technicians are right.
Answer B is a good choice because a warped cylinder head mounting surface on an engine block may cause coolant and combustion leaks. Yet, it is wrong because both technicians are right.
Answer C is correct because both technicians are right.
Answer D is wrong because both technicians are right.

Question #25
Answer A is wrong because cylinder taper is the difference between the cylinder diameter at the top of the ring travel compared to cylinder diameter at the bottom of the ring travel, not the center as stated.
Answer B is wrong because cylinder out-of-round is the difference between axial cylinder bore diameter at two 180 degree points, not the top of the ring travel, compared to the thrust cylinder bore diameter at the bottom of the ring travel.
Answer C is wrong because both technicians are wrong.
Answer D is correct because neither technician is right.

Question #26
Answer A is wrong because the skirt never meets the top of the cylinder.
Answer B is wrong because the cylinder ridge would not affect the piston pin.
Answer C is wrong because the connecting rod bearings would not be affected.
Answer D is correct because the piston ring lands may be broken.

Question #27
Answer A is correct because a normal wear pattern will have a slightly off center with wider wear pattern at the nose than at the heel.
Answer B is wrong because the normal wear pattern would NOT extend to the edge of the lobe.
Answer C is wrong because only one technician is right.
Answer D is wrong because one technician is right.

Question #28
Answer A is a good choice because a nick on any of the crankshaft journals that can be caught with a fingernail has to be serviced. Yet, it is wrong because both technicians are right.
Answer B is a good choice because crankshafts can be checked for cracks magnetically by using magnetic particle inspection. Yet, it is wrong because both technicians are right.
Answer C is correct because both technicians are right.
Answer D is wrong because both technicians are right.

Question #29
Answer A is correct because it is near impossible to determine mileage on bearing inspection, therefore the exception.
Answer B is wrong because you can determine crankshaft misalignment from bearing inspection.
Answer C is wrong because you can determine lack of lubrication during bearing inspection.
Answer D is wrong because you can determine metal-to-metal contact during bearing inspection.

Question #30
Answer A is wrong because a bent connecting rod as shown will cause uneven connecting rod bearing wear.
Answer B is correct and the exception, because only a bent crankshaft could cause main bearing wear.
Answer C is wrong because a bent connecting rod as shown will cause wear on the piston pin.
Answer D is wrong because a bent connecting rod as shown will cause a very specific cylinder wall wear pattern.

Question #31
Answer A is wrong because one does not use the tool shown in the figure to widen the piston ring grooves.
Answer B is wrong because one does not use the tool shown in the figure to deepen the piston ring grooves.
Answer C is correct because one uses the tool shown in the figure to remove and replace the piston rings.
Answer D is wrong because one does not use the tool shown in the figure to remove carbon from the piston ring grooves.

Question #32
Answer A is wrong because a lack of lubrication is also a cause for scoring and seizure but not the most likely.
Answer B is correct because overheating of the piston is the most likely reason for piston scoring.
Answer C is wrong because preignition is also a cause for scoring and seizure but not the most likely.
Answer D is wrong because over speeding is a potential cause for scoring and seizure but not the most likely.

Question #33
Answer A is wrong because if the piston ring groove gap has exceeded specifications, you replace the piston not the piston ring.
Answer B is correct because excessive piston ring clearance can cause a piston ring to break.
Answer C is wrong because only one technician is right.
Answer D is wrong because one technician is right.

Question #34
Answer A is wrong because you position the ring at the bottom of ring travel not the top.
Answer B is correct because the two compression rings are interchangeable on most pistons.
Answer C is wrong because only one technician is right.
Answer D is wrong because one technician is right.

Question #35
Answer A is correct because the vibration damper counterbalances the back-and-forth-twisting motion of the crankshaft each time the cylinder fires.
Answer B is wrong because service sleeves can be used to repair the hub.
Answer C is wrong because only one technician is right.
Answer D is wrong because one technician is right.

Question #36
Answer A is wrong because worn camshaft bearings will cause low oil pressure.
Answer B is wrong because worn crankshaft bearings will cause low oil pressure.
Answer C is wrong because weak oil pressure regulator spring tension can cause low oil pressure.
Answer D is correct because restricted pushrod oil passages will not cause low oil pressure.

Question #37
Answer A is wrong because oil pump gear lockup will cause drive failure.
Answer B is wrong because if the oil pump drive rod fails, the drive fails.
Answer C is correct because excessive engine speed will not cause an oil pump drive failure and is therefore the least likely cause.
Answer D is wrong because a defective oil pressure relief valve stuck will cause excessive stress on the drive shaft and it could fail.

Question #38
Answer A is correct because a loose alternator belt may cause a discharged battery.
Answer B is wrong because a loose alternator belt will not cause a squealing noise while decelerating, only on acceleration.
Answer C is wrong because a loose alternator belt will not cause a damaged alternator bearing.
Answer D is wrong because a loose alternator belt will not cause engine overheating.

Appendices Explanations to the Answers for the Sample Test Section 5 103

Question #39
Answer A is wrong because the tool as shown in the figure is used to check for heater core leaks, not cooling system leaks. You would attach the same tool with a different adapter at the radiator cap opening.
Answer B is wrong because the tool as shown in the figure is used to check for heater core leaks, not radiator cap pressure relief valve. You would attach this tool with an adapter directly to the cap.
Answer C is wrong because the tool as shown in the figure is used to check for heater core leaks. A hydrometer is used to check coolant specific gravity.
Answer D is correct because the tool as shown in the figure is used to check for heater core leaks.

Question #40
Answer A is wrong because water pumps do not work intermittently. They either pump or don't pump.
Answer B is correct because if the thermostat sometimes sticks, it could cause this condition.
Answer C is wrong because only one technician is right.
Answer D is wrong because one technician is right.

Question #41
Answer A is a good choice because to properly flush the radiator, reverse flushing should be performed to dislodge deposits. Yet, it is wrong because both technicians are right.
Answer B is a good choice because a flush involves draining the coolant and refilling the system with new antifreeze. Yet, it is wrong because both technicians are right.
Answer C is correct because both technicians are right.
Answer D is wrong because both technicians are right.

Question #42
Answer A is wrong because it is a rare occurrence for the pump impeller to deteriorate from corrosion.
Answer B is wrong because a water pump to engine block gasket failure is not that common of a failure.
Answer C is wrong because hub separation is rare.
Answer D is correct because lip seal failure is the most common reason for pump replacement.

Question #43
Answer A is a good choice because removing the lower hose first might make a mess, but the radiator drain petcock will not have to be opened. Yet, it is wrong because both technicians are right.
Answer B is a good choice because the first step to removing any radiator is to disconnect the negative battery cable. Yet, it is wrong because both technicians are right.
Answer C is correct because both technicians are right.
Answer D is wrong because both technicians are right.

Question #44
Answer A is wrong because you do not replace a viscous fan clutch just for a leak, it must fail a performance test.
Answer B is correct because all types of fan blades should be inspected for stress fractures.
Answer C is wrong because only one technician is right.
Answer D is wrong because one technician is right.

Question #45
Answer A is a good choice because the ignition module has malfunctioned and could be the cause. Yet, it is wrong because both technicians are right.
Answer B is a good choice because the hall effect switch in the distributor has lost its polarity and needs replacement. Yet, it is wrong because both technicians are right.
Answer C is correct because both technicians are right.
Answer D is wrong because both technicians are right.

Question #46
Answer A is wrong because the meter has an infinite reading between ground and the pickup coil. The leads are connected incorrectly for a ground check.
Answer B is wrong because a low resistance, not infinity, indicates a short in the pickup coil winding.
Answer C is wrong because this is not how to check for a ground.
Answer D is correct because resistance infinity on the two pickup coil leads indicates open. On the GM pickup coil shown, the coil resistance should be in the 500 to 1500 ohm range.

Question #47
Answer A is wrong because a cracked distributor cap is sensitive to moisture egress.
Answer B is wrong because deteriorated ignition wires are sensitive to moisture egress.
Answer C is wrong because a broken ignition coil core is sensitive to moisture egress.
Answer D is correct because a loose distributor mounting has no effect on moisture in the system. Therefore it is the exception.

Question #48
Answer A is wrong because not all faulty coils will leak oil when they fail.
Answer B is correct because ignition coils are part of both the primary and secondary ignition circuits.
Answer C is wrong because only one technician is right.
Answer D is wrong because one technician is right.

Question #49
Answer A is wrong because an ignition switch failure would affect all of the plugs not just one.
Answer B is wrong because an ignition coil failure would affect all of the plugs not just one.
Answer C is wrong because a Hall effect switch failure would affect all of the plugs not just one.
Answer D is correct because a failure of one or more of the spark plug secondary wires is the cause.

Question #50
Answer A is wrong because there are no scan tools that specifically check the coil, unless it is included in the OBDII diagnostic parameters.
Answer B is correct because you use the DMM to make Ohmic tests on the coil.
Answer C is wrong because you can use a scope, yet it is not the most commonly used tool.
Answer D is wrong because a test light cannot be used to test a coil.

Question #51
Answer A is a good choice because the underhood emission label has specific instructions for setting the timing. Yet, it is wrong because both technicians are right.
Answer B is a good choice because on non-feedback style carburetors, the engine idle speed must be verified before the timing can be adjusted. Yet, it is wrong because both technicians are right.
Answer C is correct because both technicians are right.
Answer D is wrong because both technicians are right.

Question #52
Answer A is wrong because an increase in point gap decreases not increases the cam dwell reading.
Answer B is correct because point dwell must be long enough to allow the magnetic field to build up in the coil.
Answer C is wrong because only one technician is right.
Answer D is wrong because one technician is right.

Question #53
Answer A is wrong because the tester has no capability to analyze the primary circuit.
Answer B is wrong because the tester is not a voltmeter.
Answer C is wrong because the tester has no capability to analyze the secondary circuit.
Answer D is correct because the ignition module tester checks the ability of the module to turn on and off.

Question #54
Answer A is a good choice because the fuel level should be checked first on a no-start condition. Yet, it is wrong because both technicians are right.
Answer B is a good choice because the injector pulse width should be checked. Yet, it is wrong because both technicians are right.
Answer C is correct because both technicians are right.
Answer D is wrong because both technicians are right.

Question #55
Answer A is wrong because if there is no fuel leaking from the fuel pump, then you still need to test it for internal leaks.
Answer B is correct because the fuel pump could be leaking internally, and the engine oil level should be checked to help indicate a problem.
Answer C is wrong because only one technician is right.
Answer D is wrong because one technician is right.

Question #56
Answer A is wrong because this is one of the first components to be disconnected.
Answer B is wrong because this is one of the first components to be disconnected.
Answer C is correct because the fuel supply line is the last component to be disconnected.
Answer D is wrong because this is one of the first components to be disconnected.

Question #57
Answer A is correct because the solution used to clean carburetors is corrosive and the plastic components should be removed beforehand.
Answer B is wrong because there is no special coating on feedback carburetor throttle plates.
Answer C is wrong because only one technician is right.
Answer D is wrong because one technician is right.

Question #58
Answer A is wrong because the PCV valve supply hose is connected to the throttle body base plate or intake manifold vacuum not throttle body vacuum.
Answer B is wrong because you cannot reuse the throttle body gasket.
Answer C is wrong because both technicians are wrong.
Answer D is correct because both technicians are wrong.

Question #59
Answer A is wrong because any type of sealer cannot be substituted for the intake gaskets.
Answer B is correct because intake manifold gaskets can be bought in a kit with all other gaskets in the intake system.
Answer C is wrong because only one technician is right.
Answer D is wrong because one technician is right.

Question #60
Answer A is wrong because there is no device that limits the amount of air into the engine.
Answer B is wrong because the PCV system does not need blowby vapor to operate.
Answer C is correct because the automatic choke lowers the pressure in the carburetor venturi and enriches the mixture thus providing the best engine performance in cold weather. Cold air does not absorb much moisture so the increased fuel allows the engine to start easier in cold weather.
Answer D is wrong because providing raw fuel is not the function of the choke.

Question #61
Answer A is correct because in closed loop operation is when the engine is at operating temperature and the PCM uses the information from the oxygen sensor to control the air-fuel ratio.
Answer B is wrong because injector dwell time is measured in milliseconds.
Answer C is wrong because only one technician is right.
Answer D is wrong because one technician is right.

Question #62
Answer A is wrong because the first sensor value to change is the throttle position.
Answer B is wrong because this is not the first sensor condition to change.
Answer C is wrong because this is the second sensor value to change.
Answer D is correct because the throttle position sensor (TPS) is the first sensor to change during acceleration.

Question #63
Answer A is wrong because fuel-injected vehicles need fuel filter replacement on regular intervals.
Answer B is correct because a shop light can be used to determine how dirty the element is.
Answer C is wrong because only one technician is right.
Answer D is wrong because one technician is right.

Question #64
Answer A is a good choice because the fuel pressure regulator senses engine vacuum. Yet, it is wrong because both technicians are right.
Answer B is a good choice because the fuel pressure regulator controls the amount of fuel that is returned to the tank. Yet, it is wrong because both technicians are right.
Answer C is correct because both technicians are right.
Answer D is wrong because both technicians are right.

Question #65
Answer A is wrong because stainless steel manifolds are usually more costly.
Answer B is wrong because they are lighter in weight.
Answer C is wrong because they do not produce higher emissions.
Answer D is correct because they are lighter in weight.

Question #66
Answer A is correct because a battery can be charged at any rate as long as the electrolyte, does not boil and the temperature does not exceed 120°F (50°C).
Answer B is wrong because fast charging is not as complete as charging a battery at a slower rate.
Answer C is wrong because only one technician is right.
Answer D is wrong because one technician is right.

Question #67
Answer A is wrong because the terminal should be removed with a puller.
Answer B is wrong because you always remove the battery cable first.
Answer C is correct because you never remove the positive battery cable before the negative cable.
Answer D is wrong because a baking soda and water solution can be used to clean the battery.

Question #68
Answer A is wrong because there is no alternator current test to perform.
Answer B is wrong because the amp pickup lead would have to be placed on the starter cable.
Answer C is correct because the battery load test is shown in the figure.
Answer D is wrong because grounding circuit connections are checked with an ohmmeter.

Question #69
Answer A is wrong because when jump starting a vehicle with a dead battery, you never place rags over the battery to be boosted.
Answer B is wrong because you connect the negative cables last to prevent sparks.
Answer C is wrong because both technicians are wrong.
Answer D is correct because neither technician is right.

Question #70
Answer A is a good choice because battery terminals can be treated with petroleum jelly to help prevent corrosion. Yet, it is wrong because both technicians are right.
Answer B is a good choice because protective pads can be placed on the terminals to prevent corrosion. Yet, it is wrong because both technicians are right.
Answer C is correct because both technicians are right.
Answer D is wrong because both technicians are right.

Question #71
Answer A is wrong because starter motor current draw can only be tested on the bench.
Answer B is correct because the starter motor current draw test is shown.
Answer C is wrong because the amp pickup lead is on the starter cable.
Answer D is wrong because the resistance in the circuit is not being checked.

Question #72
Answer A is correct because a starting and charging system tester can be used to test free spinning current draw tests.
Answer B is wrong because you do not have to remove the solenoid to do the free spinning current draw test.
Answer C is wrong because only one technician is right.
Answer D is wrong because one technician is right.

Question #73
Answer A is wrong because EGR operates on ported venturi vacuum systems.
Answer B is wrong because EGR operates on ported vacuum.
Answer C is wrong because EGR valves can be mounted on the intake.
Answer D is correct because they do not operate at wide open throttle.

Question #74
Answer A is wrong because the coolant temperature override switch limits EGR valve operation until the engine warms up to operating temperature.
Answer B is correct because the EVR solenoid converts the electrical signals from the PCM into a mechanical action, which directs vacuum to the EGR valve.
Answer C is wrong because only one technician is right.
Answer D is wrong because one technician is right.

Question #75
Answer A is correct because the air pump drive belt is the first area you check.
Answer B is wrong because the air-injection plumbing is not the most logical component to be inspected first.
Answer C is wrong because this is not the easiest thing to perform first.
Answer D is wrong because it is easier to inspect the belt first.

Question #76
Answer A is wrong because you cannot rebuild the air pump.
Answer B is correct because you should never pry on the housing while tightening the drive belt.
Answer C is wrong because only one technician is right.
Answer D is wrong because one technician is right.

Question #77
Answer A is a good choice because the gulp valve is used to prevent engine backfire. Yet, it is wrong because both technicians are right.
Answer B is a good choice because the gulp valve has largely been replaced by the diverter valve. Yet, it is wrong because both technicians are right.
Answer C is correct because both technicians are right.
Answer D is wrong because both technicians are right.

Question #78
Answer A is correct because preformed air-injection hose is the correct material for the repair.
Answer B is wrong because heater hose is not suitable for air-injection hose.
Answer C is wrong because fuel supply hose is too small in diameter.
Answer D is wrong because heater hose is not the correct diameter.

Question #79
Answer A is wrong because oil consumption is not a direct indication of PCV faults.
Answer B is wrong because normal HC and CO emission readings would indicate that the PCV is operating normally.
Answer C is wrong because normal fuel consumption would indicate that the PCV is operating normally.
Answer D is correct because excessive blowby is a direct indication of a PCV failure.

Question #80
Answer A is wrong because all three-way catalytic (TWC) converters are not interchangeable.
Answer B is wrong because an overly rich air-fuel mixture will affect the catalytic converter.
Answer C is wrong because both technicians are wrong.
Answer D is correct because neither technician is right.

Question #81
Answer A is a good choice because low voltage at the fuel pump will cause a severe surging problem only at 55 mph. Yet, it is wrong because both technicians are right.
Answer B is a good choice because the inertia switch may have high resistance and will cause a severe surging problem only at 55 mph. Yet, it is wrong because both technicians are right.
Answer C is correct because both technicians are right.
Answer D is wrong because both technicians are right.

Question #82
Answer A is correct because you check the ECT sensor signal to the PCM. When diagnosing an EVAP system with a scan tool, the PCM never provides the on command to the EVAP solenoid at any engine or vehicle speed.
Answer B is wrong because the vacuum hoses from the intake to the EVAP canister will have no effect on the ECT signal.
Answer C is wrong because only one technician is right.
Answer D is wrong because one technician is right.

Question #83
Answer A is a good choice because a defective ECT sensor may cause hard cold starting. Yet, it is wrong because both technicians are right.
Answer B is a good choice because a defective ECT sensor may cause improper emissions. Yet, it is wrong because both technicians are right.
Answer C is correct because both technicians are right.
Answer D is wrong because both technicians are right.

Question #84
Answer A is correct because the sensor can be checked with the meter on the resistance scale or the voltage scale.
Answer B is wrong because if the PCM does not set a throttle position sensor fault code, there may still be a problem with the sensor.
Answer C is wrong because only one technician is right.
Answer D is wrong because one technician is right.

Question #85
Answer A is a good choice because most manufacturers provide technical service information on compact discs (CD). Yet, it is wrong because both technicians are right.
Answer B is a good choice because if available, service manuals from the vehicle's manufacturer have more detailed information regarding procedures and diagnostic information. Yet, it is wrong because both technicians are right.
Answer C is correct because both technicians are right.
Answer D is wrong because both technicians are right.

Question #86
Answer A is wrong because circuit protection devices open when the current in the circuit falls above a predetermined level.
Answer B is correct because excessive current results from a decrease in the circuit's resistance.
Answer C is wrong because only one technician is right.
Answer D is wrong because one technician is right.

Answers for the Additional Test Questions Section 6

1.	A	30.	B	59.	B	88.	D
2.	C	31.	C	60.	C	89.	C
3.	B	32.	D	61.	D	90.	D
4.	C	33.	C	62.	A	91.	C
5.	A	34.	A	63.	B	92.	B
6.	D	35.	B	64.	C	93.	C
7.	C	36.	D	65.	B	94.	C
8.	B	37.	A	66.	A	95.	A
9.	B	38.	C	67.	C	96.	B
10.	C	39.	C	68.	D	97.	B
11.	B	40.	D	69.	B	98.	B
12.	C	41.	A	70.	D	99.	C
13.	A	42.	B	71.	B	100.	D
14.	C	43.	A	72.	A	101.	B
15.	B	44.	D	73.	D	102.	D
16.	D	45.	D	74.	D	103.	D
17.	C	46.	A	75.	C	104.	C
18.	B	47.	B	76.	B	105.	B
19.	A	48.	D	77.	C	106.	A
20.	B	49.	B	78.	C	107.	C
21.	D	50.	A	79.	D	108.	B
22.	B	51.	B	80.	A	109.	C
23.	C	52.	C	81.	C	110.	D
24.	C	53.	C	82.	C	111.	A
25.	A	54.	C	83.	C	112.	C
26.	C	55.	D	84.	B	113.	D
27.	C	56.	C	85.	C		
28.	C	57.	A	86.	D		
29.	D	58.	B	87.	D		

Explanations to the Answers for the Additional Test Questions Section 6

Question #1
Answer A is correct because after the repair you take the vehicle on a road test.
Answer B is wrong because you top off the fluids only if the customer asks for it.
Answer C is wrong because this should be done before the vehicle enters the shop.
Answer D is wrong because underhood detailing is not needed if the customer does not ask for it.

Question #2
Answer A is a good choice because with no visible external signs, the engine may have a leaking head gasket. Yet, it is wrong because both technicians are right.
Answer B is a good choice because the heater core may be leaking. Yet, it is wrong because both technicians are right.
Answer C is correct because both technicians are right.
Answer D is wrong because both technicians are right.

Question #3
Answer A is wrong because piston noise would diminish after the engine was at operating temperature.
Answer B is correct because loose flywheel bolts can cause this thumping noise.
Answer C is wrong because if the main bearings were in question, the oil pressure would indicate there was a problem.
Answer D is wrong because, again, oil pressure would indicate there was a problem.

Question #4
Answer A is wrong because engine coolant cannot be tested for these two things.
Answer B is wrong because lead content and the viscosity are not of concern when discussing coolant.
Answer C is correct because you can test coolant for acidity and freezing point.
Answer D is wrong because water pump condition cannot be determined by the condition of the coolant.

Question #5
Answer A is correct because a burned exhaust valve will cause a steady puff noise coming from the exhaust with the engine idling.
Answer B is wrong because the fuel pressure regulator would not cause an engine mechanical problem from the exhaust.
Answer C is wrong because a restricted fuel line could not cause an exhaust puff.
Answer D is wrong because fuel pump check valves have nothing to do with engine mechanical problems.

Question #6
Answer A is wrong because late ignition timing would not cause fluctuations.
Answer B is wrong because intake manifold leaks will cause a low vacuum reading.
Answer C is wrong because a restricted exhaust system will cause a high vacuum reading.
Answer D is correct because sticking valve stems and guides cause a vacuum gauge needle fluctuation in the 15 to 20 in. Hg range.

Question #7
Answer A is wrong because the battery should already be charged.
Answer B is wrong because it does not make a difference if the vehicle is warmed up or not.
Answer C is correct because the first step in doing a compression test is to disable the ignition and the fuel system.
Answer D is wrong because holding the accelerator is not the first thing to be performed.

Question #8
Answer A is wrong because a leaking intake valve would not pass air into the crankcase, only through the throttle.
Answer B is correct because if the piston rings in that cylinder are worn, air would pass into the crankcase and out the PCV.
Answer C is wrong because only one technician is right.
Answer D is wrong because one technician is right.

Question #9
Answer A is wrong because a fouled spark plug would not fire at all and on a scope show a very low firing line.
Answer B is correct because circuit gap or firing line voltage is the voltage required to fire all the spark plugs in the secondary circuit.
Answer C is wrong because only one technician is right.
Answer D is wrong because one technician is right.

Question #10
Answer A is a good choice because the cylinder is filled with regulated shop air. Yet, it is wrong because both technicians are right.
Answer B is a good choice because both intake and exhaust valves must be closed. Yet, it is wrong because both technicians are right.
Answer C is correct because both technicians are right.
Answer D is wrong because both technicians are right.

Question #11
Answer A is wrong because valve seats are replaced separately.
Answer B is correct because seat run out is a measure of how circular the valve seat is in relation to the valve guide.
Answer C is wrong because only one technician is right.
Answer D is wrong because one technician is right.

Question #12
Answer A is a good choice because the three high stress areas of the rocker arm are the pivot, pushrod contact, and valve stem contact surfaces. Yet, it is wrong because both technicians are right.
Answer B is a good choice because if valve stem wear is not centered in the rocker arm, it may be caused by a bent stud. Yet, it is wrong because both technicians are right.
Answer C is correct because both technicians are right.
Answer D is wrong because both technicians are right.

Question #13
Answer A is correct because the valve stem must be measured at the top, middle, and near the fillet.
Answer B is wrong because modern engines do not use tapered stems.
Answer C is wrong because only one technician is right.
Answer D is wrong because one technician is right.

Question #14
Answer A is a good choice because the valve is stroked across the stone as the valve is being fed in. Yet, it is wrong because both technicians are right.
Answer B is a good choice because after the last pass is complete you back the valve away from the stone. Yet, it is wrong because both technicians are right.
Answer C is correct because both technicians are right.
Answer D is wrong because both technicians are right.

Question #15
Answer A is wrong because the amount of material removed from the valve face and the shim size have nothing to do with each other.
Answer B is correct because valve stem tip height must be correct for proper rocker arm geometry.
Answer C is wrong because only one technician is right.
Answer D is wrong because one technician is right.

Question #16
Answer A is wrong because after the valve springs are installed, you never attempt to turn the valve spring by hand.
Answer B is wrong because the spring rotation is not an indication of proper valve spring installation. The valve spring will probably not rotate.
Answer C is wrong because both technicians are wrong.
Answer D is correct because neither technician is right.

Question #17
Answer A is a good choice because a worn-out valve guide will provide an inaccurate valve seat run out measurement. Yet, it is wrong because both technicians are right.
Answer B is a good choice because the valve seat to valve face contact area provides a path for heat from the valve head to dissipate. Yet, it is wrong because both technicians are right.
Answer C is correct because both technicians are right.
Answer D is wrong because both technicians are right.

Question #18
Answer A is wrong because you install the serrated side toward the head, not the spring.
Answer B is correct because to determine required shim thickness, the measured distance is subtracted from the specifications.
Answer C is wrong because only one technician is right.
Answer D is wrong because one technician is right.

Question #19
Answer A is correct because the pushrods can be checked for straightness by rolling them on a known flat surface.
Answer B is wrong because bent pushrods are not always visible to the eye.
Answer C is wrong because only one technician is right.
Answer D is wrong because one technician is right.

Question #20
Answer A is wrong because priming new lifters only requires soaking them in oil.
Answer B is correct because the tool is used to determine lifter leak down.
Answer C is wrong because only one technician is right.
Answer D is wrong because one technician is right.

Question #21
Answer A is wrong because the leftover residue from the carburetor cleaner can affect the gasket surface.
Answer B is wrong because RTV should be placed completely around the bolt hole.
Answer C is wrong because RTV will set up after 10 to 15 minutes.
Answer D is correct because the RTV bead should be 1/8 in. (3 mm) in width.

Question #22
Answer A is wrong because block fatigue can cause cracks.
Answer B is correct because loss of oil pressure will cause seizure but not cracks.
Answer C is wrong because impact damage will cause cracks.
Answer D is wrong because detonation will cause cracks.

Question #23
Answer A is a good choice because taper in the bore causes the ring end gaps to change while the piston moves in the bore. Yet, it is wrong because both technicians are right.
Answer B is a good choice because a cylinder bore gauge can be used to determine taper. Yet, it is wrong because both technicians are right.
Answer C is correct because both technicians are right.
Answer D is wrong because both technicians are right.

Question #24
Answer A is a good choice because on a valve clearance check the feeler gauge should slide with slight resistance. Yet, it is wrong because both technicians are right.
Answer B is also a good choice because intake valve clearance is less than the clearance for the exhaust valve due to thermal expansion. Yet, it is wrong because both technicians are right.
Answer C is correct because both technicians are right.
Answer D is wrong because both technicians are right.

Question #25
Answer A is correct because you place an oiled rag in the cylinder to catch any metal shavings.
Answer B is wrong because you never use sandpaper on the cylinder walls.
Answer C is wrong because a step of 0.050 in. (1.2 mm) is excessive and the ridge needs to be removed.
Answer D is wrong because you only wash the cylinders after the ridges have been removed.

Question #26
Answer A is a good choice because Plastigage® can be placed between the camshaft journal and the bearing cap. Yet, it is wrong because both technicians are right.
Answer B is a good choice because overhead camshaft bearings are made of aluminum and are inserts. Yet, it is wrong because both technicians are right.
Answer C is correct because both technicians are right.
Answer D is wrong because both technicians are right.

Question #27
Answer A is a good choice because metal burrs on the crankshaft flange may cause excessive wear on the flexplate gear teeth. Yet, it is wrong because both technicians are right.
Answer B is a good choice because metal burrs may cause improper torque converter to transmission alignment. Yet, it is wrong because both technicians are right.
Answer C is correct because both technicians are right.
Answer D is wrong because both technicians are right.

Question #28
Answer A is wrong because if a scratch is deep enough to catch with a fingernail, the crankshaft needs machining.
Answer B is wrong because there is no special tool shown in the figure.
Answer C is correct because the technician is measuring the thickness of the crushed Plastigage® in the figure.
Answer D is wrong because to determine if the crankshaft has been machined, the bearing insert will indicate that it is oversized.

Question #29
Answer A is wrong because the fixture shown is used to bore the connecting pin bushing or big end.
Answer B is wrong because the fixture shown is used to bore the connecting pin bushing or big end.
Answer C is wrong because both technicians are wrong.
Answer D is correct because neither technician is right.

Question #30
Answer A is wrong because you never drive out a piston pin with a hammer and drift.
Answer B is correct because removing the pin requires a press and special adapters.
Answer C is wrong because only one technician is right.
Answer D is wrong because one technician is right.

Question #31
Answer A is a good choice because pistons are cam ground. Yet, it is wrong because both technicians are right.
Answer B is a good choice because pistons should be measured across the thrust surface of the skirt centerline of the piston pin. Yet, it is wrong because both technicians are right.
Answer C is correct because both technicians are right.
Answer D is wrong because both technicians are right.

Question #32
Answer A is wrong because if the groove opening is larger than specifications, a new piston should be installed.
Answer B is wrong because excessive piston ring side clearance is not recommended, but it will harm the rings.
Answer C is wrong because both technicians are wrong.
Answer D is correct because neither technician is right.

Question #33
Answer A is wrong because this is not an approved method for fixing a balancer.
Answer B is wrong because this will melt the rubber insulator.
Answer C is correct because if the outer ring has slipped, you replace the harmonic balancer.
Answer D is wrong because outer rings are not available separately.

Question #34
Answer A is correct because inner rotor diameter is not something you normally measure.
Answer B is wrong because you do measure clearance between the rotors.
Answer C is wrong because you do measure inner and outer rotor thickness.
Answer D is wrong because you do measure outer rotor to housing clearance.

Question #35
Answer A is wrong because oil pumps are cheap compared to the labor which it would take to rebuild it.
Answer B is correct because it is a good practice to replace an oil pump when rebuilding an engine.
Answer C is wrong because only one technician is right.
Answer D is wrong because one technician is right.

Question #36
Answer A is wrong because you can use a belt tension gauge.
Answer B is wrong because you can measure the amount of belt deflection.
Answer C is wrong because you can visually see if the belt is contacting the bottom of the pulley.
Answer D is correct because you do not measure the length of the belt compared to a new one.

Question #37
Answer A is correct because radiator pressure caps can be tested with a special adapter.
Answer B is wrong because pressure testing the coolant system only measures the system integrity. The thermostat must be tested for opening and closing.
Answer C is wrong because only one technician is right.
Answer D is wrong because one technician is right.

Question #38
Answer A is a good choice because you fill half the system capacity with pure antifreeze and the remaining amount with water. Yet, it is wrong because both technicians are right.
Answer B is a good choice because you use a 50/50 mixture of water to coolant mixture to fill the cooling system. Yet, it is wrong because both technicians are right.
Answer C is correct because both technicians are right.
Answer D is wrong because both technicians are right.

Question #39
Answer A is a good choice because a defective water pump bearing may cause a growling noise when the engine is idling. Yet, it is wrong because both technicians are right.
Answer B is a good choice because the water pump bearing may be ruined by coolant leaking past the pump seal. Yet, it is wrong because both technicians are right.
Answer C is correct because both technicians are right.
Answer D is wrong because both technicians are right.

Question #40
Answer A is wrong because there is no such radiator as a downdraft.
Answer B is wrong because there is no such radiator as an updraft.
Answer C is wrong because this is not the type of radiator shown.
Answer D is correct because a downflow radiator is shown.

Question #41
Answer A is correct because a stuck closed vacuum valve would cause a collapsed upper radiator hose after the engine is shut off. With the vacuum valve closed, no atmospheric pressure would enter the system. As the coolant cools the volume increases, the pressure decreases, and normally the vacuum valve allows air to balance this pressure. It is does not, the outside pressure would collapse the hose because there is higher pressure outside than inside.
Answer B is wrong because a faulty vacuum valve would not cause the system to overpressurize.
Answer C is wrong because if the engine is overheating under a load, it is more likely that the ignition timing is incorrect.
Answer D is wrong because this would indicate cooling fan problems.

Question #42
Answer A is wrong because intake manifold leaks commonly cause misfires.
Answer B is correct because an exhaust manifold leak is the least likely cause of a misfire.
Answer C is wrong because spark plugs are often the cause of misfires.
Answer D is wrong because a defective coil can cause misfires.

Question #43
Answer A is correct because with a 12-volt test lamp connected to the tach terminal, if the light flutters, the primary ignition system is functioning properly.
Answer B is wrong because the pickup coil can be tested with an ohmmeter.
Answer C is wrong because only one technician is right.
Answer D is wrong because one technician is right.

Question #44
Answer A is wrong because you do not check a mechanical advance to ensure that it moves when vacuum is applied.
Answer B is wrong because you do not check a mechanical advance for cracks in the diaphragm.
Answer C is wrong because you do not check a mechanical advance for brittle vacuum supply hose.
Answer D is correct because you check a mechanical advance for broken or missing counterweight springs.

Question #45
Answer A is wrong because secondary ignition circuit does not start at the ignition switch and follow through to the spark plug.
Answer B is wrong because the primary circuit is not from the coil terminal through the distributor cap to the spark plugs.
Answer C is wrong because both technicians are wrong.
Answer D is correct because neither technician is right.

Question #46
Answer A is correct because the primary windings should have a resistance of about 0.5 to 2.0 ohms.
Answer B is wrong because any infinity reading indicates an open and reason for rejection.
Answer C is wrong because only one technician is right.
Answer D is wrong because one technician is right.

Question #47
Answer A is wrong because timing lights cannot be used as trouble lights.
Answer B is correct because this timing light can check spark advance.
Answer C is wrong because timing lights cannot be used as oscilloscopes.
Answer D is wrong because all timing checks occur on the secondary side.

Question #48
Answer A is wrong because only an internal pickup coil winding failure can cause open in the winding.
Answer B is wrong because no known pickup coils have an infinity reading or they could not generate a voltage.
Answer C is wrong because both technicians are wrong.
Answer D is correct because neither technician is correct.

Question #49
Answer A is wrong because ignition module testers are manufacturer specific.
Answer B is correct because each vehicle manufacturer has their own ignition module tester.
Answer C is wrong because only one technician is right.
Answer D is wrong because one technician is right.

Question #50
Answer A is correct because fuel injectors are supplied with 12-volt power at all times.
Answer B is wrong because a no-start concern can be more than just an electrical malfunction.
Answer C is wrong because only one technician is right.
Answer D is wrong because one technician is right.

Question #51
Answer A is wrong because fuel pumps are located in the tank to cool the pump.
Answer B is correct because the fuel pumps run for 5 to 10 seconds during the ignition on cycle to pressurize the fuel rail during startup.
Answer C is wrong because only one technician is right.
Answer D is wrong because one technician is right.

Question #52
Answer A is wrong because the fuel line is the last component to be removed.
Answer B is wrong because you never use open flame around the carburetor.
Answer C is correct because you label all connections so that reassembly will not be confusing.
Answer D is wrong because asking another technician will not fix the vehicle.

Question #53
Answer A is a good choice because this is a good time to check for signs of obvious damage. Yet, it is wrong because both technicians are right.
Answer B is a good choice because the subassemblies should be inspected at the same time. Yet, it is wrong because both technicians are right.
Answer C is correct because both technicians are right.
Answer D is wrong because both technicians are right.

Question #54
Answer A is wrong because the coating has nothing to do with engine backfiring.
Answer B is wrong because the coating does not affect the airflow.
Answer C is correct because the coating prevents carbon buildup on the throttle plate.
Answer D is wrong because the special coating cannot increase the incoming airflow.

Question #55
Answer A is wrong because it's a good practice to replace all the intake manifold gaskets at the same time.
Answer B is wrong because the replacement gasket should be made of the same material as the original.
Answer C is wrong because all the gaskets in the intake system do come together in a kit.
Answer D is correct because cracked intake manifold gaskets cannot be repaired with silicone.

Question #56
Answer A is a good choice because the choke must be inspected for proper operation before an adjustment can be made. Yet, it is wrong because both technicians are right.
Answer B is a good choice because the choke linkage must be free from dirt and debris. Yet, it is wrong because both technicians are right.
Answer C is correct because both technicians are right.
Answer D is wrong because both technicians are right.

Question #57
Answer A is correct because the PCM strategy monitors various sensors in order of importance.
Answer B is wrong because if the engine coolant temperature drops below a specified temperature, the PCM strategy increases, not decreases, the injector pulse width.
Answer C is wrong because only one technician is right.
Answer D is wrong because one technician is right.

Question #58
Answer A is wrong because the PCM does not use anti-lock information to determine if the vehicle is decelerating.
Answer B is correct because the throttle position sensor determines deceleration mode.
Answer C is wrong because the purge solenoid can give no indication of deceleration mode.
Answer D is wrong because the camshaft position sensor will not indicate deceleration mode.

Question #59
Answer A is wrong because the optimal air-fuel ratio is 14.7 to 1.
Answer B is correct because for every gallon of gasoline that the engine burns, the engine uses 9,000 gallons of air.
Answer C is wrong because only one technician is right.
Answer D is wrong because one technician is right.

Question #60
Answer A is wrong because a decrease in engine vacuum indicates an increase in load, which requires more fuel pressure.
Answer B is wrong because the valve closes with an increase in engine load.
Answer C is correct because when the engine vacuum decreases, the fuel pressure regulator spring pushes the valve closed to allow fuel pressure to build in the rail.
Answer D is wrong because the fuel pump is always operating, no matter what the load is.

Question #61
Answer A is wrong because cast iron manifolds are significantly heavier than stainless steel.
Answer B is wrong because cast iron manifolds do offer good resistance to change in temperature.
Answer C is wrong because both technicians are wrong.
Answer D is correct because neither technician is right.

Question #62
Answer A is correct because the solution of baking soda and water should be allowed to run over the battery into the tray.
Answer B is wrong because you never use anything but water or distilled water to fill the cells of the battery.
Answer C is wrong because only one technician is right.
Answer D is wrong because one technician is right.

Question #63
Answer A is wrong because you never let the battery temperature rise over 125°F (52°C) while charging the battery.
Answer B is correct because the slower the battery is charged with a lower amp level, the more complete the charge.
Answer C is wrong because only one technician is right.
Answer D is wrong because one technician is right.

Question #64
Answer A is wrong because the load should be half that of the cold cranking amps.
Answer B is wrong because the load should be half that of the cold cranking amps.
Answer C is correct because the load should be half that of the cold cranking amps and 262 is half of the cold cranking amps.
Answer D is wrong because the load should be half that of the cold cranking amps.

Question #65
Answer A is wrong because you connect the negative battery cable last, not first as stated on the discharged battery.
Answer B is correct because all the accessories must be turned off to prevent electrical damage.
Answer C is wrong because only one technician is right.
Answer D is wrong because one technician is right.

Question #66
Answer A is correct because you use a wire brush to remove corrosion on the battery terminals.
Answer B is wrong because the post can be cleaned using a battery terminal brush.
Answer C is wrong because only one technician is right.
Answer D is wrong because one technician is right.

Question #67
Answer A is a good choice because high resistance in the starter motor circuit can cause low cranking speed. Yet, it is wrong because both technicians are right.
Answer B is a good choice because internal engine problems can be the cause of the starter motor not functioning. Yet, it is wrong because both technicians are right.
Answer C is correct because both technicians are right.
Answer D is wrong because both technicians are right.

Question #68
Answer A is wrong because pinion gear clearance is checked using a feeler gauge.
Answer B is wrong because the starter would have to be installed in the vehicle to perform this test.
Answer C is wrong because the battery is not shown in the figure.
Answer D is correct because the starter free speed test is shown in the figure.

Question #69
Answer A is wrong because if the carbon buildup on the EGR valve cannot be cleaned with solvent, the valve should be replaced.
Answer B is correct because you do use fuel-injection cleaner to clean the passages in the EGR valve.
Answer C is wrong because only one technician is right.
Answer D is wrong because one technician is right.

Question #70
Answer A is wrong because vacuum hoses usually do cause failures with the EGR system.
Answer B is wrong because the plastic used for the EPT and EVR solenoids is not very durable and will not crack around the vacuum hose connection.
Answer C is wrong because both technicians are wrong.
Answer D is correct because neither technician is right.

Question #71
Answer A is wrong because the hoses indicated are outlet hoses.
Answer B is correct because either hose should discharge air flow out the end of the hose.
Answer C is wrong because the air pump is always positive displacement no matter what the rpm is.
Answer D is wrong because there is no air pump clutch.

Question #72
Answer A is correct because new air pumps will come with a new diverter valve.
Answer B is wrong because new air pumps come with a new pulley.
Answer C is wrong because only one technician is right.
Answer D is wrong because one technician is right.

Question #73
Answer A is wrong because if there were no vacuum present, the engine would be off or the hose clogged or restricted.
Answer B is wrong because you would not feel positive air pressure at this hose, only vacuum or nothing.
Answer C is wrong because you would never feel the presence of exhaust gas at this hose.
Answer D is correct because after disconnecting the vacuum supply hose to the bypass valve, vacuum should be present at the hose end.

Question #74
Answer A is wrong because you remove the exhaust manifold from the vehicle after you remove the air-injection manifold.
Answer B is wrong because you do not have to remove the air pump from the vehicle to remove the air-injection manifold.
Answer C is wrong because you do not use high temperature silicone in place of the gasket.
Answer D is correct because you do apply penetrating oil to the nuts before removal.

Question #75
Answer A is a good choice because regular oil change intervals prevent clogs in the PCV system. Yet, it is wrong because both technicians are right.
Answer B is a good choice because you do check the air filter for oil deposits and oil puddling. Yet, it is wrong because both technicians are right.
Answer C is correct because both technicians are right.
Answer D is wrong because both technicians are right.

Question #76
Answer A is wrong because you would remove the catalytic converter while it is cold, not at operating temperature, which is about 1100°F.
Answer B is correct because the heat shields should be reinstalled after the catalytic converter is replaced.
Answer C is wrong because only one technician is right.
Answer D is wrong because one technician is right.

Question #77
Answer A is wrong because on most OBD systems, oxygen sensor failure has no effect on cruise control operation.
Answer B is wrong because on most OBD systems, oxygen sensor failure will put the engine air-fuel ratio at a fixed ratio of 14.7:1.
Answer C is correct because oxygen sensor failure will prevent the OBD system from going to closed loop.
Answer D is wrong because oxygen sensor failure has no effect on the production of oxides of nitrogen (NO_x).

Question #78
Answer A is wrong because you do warm the engine to normal operating temperature when using a scan tool.
Answer B is wrong because you do connect the power adapter to the cigar lighter.
Answer C is correct because you do not disconnect the negative battery cable before connecting the DLC connector. Therefore, this is the exception.
Answer D is wrong because you do enter the model year, engine size, and vehicle model.

Question #79
Answer A is wrong because a scan tool is used to diagnose an engine coolant temperature sensor (ECT).
Answer B is wrong because a DMM is used to diagnose an engine coolant temperature sensor (ECT).
Answer C is wrong because you can observe resistance values at different temperatures to diagnose an engine coolant temperature sensor (ECT).
Answer D is correct because using a variable resistor to diagnose an ECT is not an approved method.

Question #80
Answer A is correct because kiloamps are not used on a DMM.
Answer B is wrong because DC volts are found on a DMM.
Answer C is wrong because AC volts are found on a DMM.
Answer D is wrong because ohms are found on a DMM.

Question #81
Answer A is a good choice because service manuals cannot anticipate all types of situations that may occur to a technician performing work on a vehicle. Yet, it is wrong because both technicians are right.
Answer B is a good choice because service manuals are published in different languages. Yet, it is wrong because both technicians are right.
Answer C is correct because both technicians are right.
Answer D is wrong because both technicians are right.

Question #82
Answer A is wrong because maxi-fuses are good for one time only.
Answer B is wrong because fusible links are good for one time only.
Answer C is correct because a circuit breaker can be used many times.
Answer D is wrong because a fuse is good for one time only.

Question #83
Answer A is a good choice because a good diagnostic strategy is to think of possible causes of the indicated problem. Yet, it is wrong because both technicians are right.
Answer B is a good choice because you check the fuel level first on driveability concerns. Yet, it is wrong because both technicians are right.
Answer C is correct because both technicians are right.
Answer D is wrong because both technicians are right.

Question #84
Answer A is wrong because gasket failure is not the only cause of oil leaks.
Answer B is correct because oil leaks can be found by adding a dye to the engine oil that is visible under ultraviolet light.
Answer C is wrong because only one technician is right.
Answer D is wrong because one technician is right.

Question #85
Answer A is wrong because freezing point cannot be determined by the pH.
Answer B is wrong because the condition of the head gasket cannot be determined without disassembly.
Answer C is correct because coolant acidity can be checked with pH strips.
Answer D is wrong because there is no way to determine the age of the coolant other than how many miles it has been used in the engine.

Question #86
Answer A is wrong because a retarded timing belt adjustment will not cause an engine metallic noise.
Answer B is wrong because a loose crankshaft dampener bolt will make a loud bottom end noise, not like valve train noise or tapping.
Answer C is wrong. A loose serpentine drive belt will not make a valve train type noise because it is made of a nonmetallic material.
Answer D is correct because lack of lubrication will collapse hydraulic lifters and make valve train noise.

Question #87
Answer A is wrong because blue colored smoke from the tailpipe usually indicates an internal oil leak.
Answer B is wrong because excessive rattling of the exhaust system would only indicate a very severe misfire.
Answer C is wrong because misfires cannot be determined by exhaust temperature.
Answer D is correct because puffing or wheezing indicates an engine misfire.

Question #88
Answer A is wrong because a fluctuating gauge needle indicates weak valve springs.
Answer B is wrong because a fluctuating gauge needle would indicate burned intake valves.
Answer C is wrong because a regular gauge fluctuating would indicate a head gasket failure.
Answer D is correct because vacuum gauge fluctuations between 7 and 20 in. Hg would indicate a burned exhaust valve.

Question #89
Answer A is a good choice because when all the cylinders are contributing equally to the engine power, all the cylinders will provide the specified rpm drop. Yet, it is wrong because both technicians are right.
Answer B is a good choice because often the reason for a cylinder not contributing to the engine power can be traced to the fuel or the ignition system. Yet, it is wrong because both technicians are right.
Answer C is correct because both technicians are right.
Answer D is wrong because both technicians are right.

Question #90
Answer A is wrong because while performing a compression test, a gradual buildup of compression with each stroke is normal and nothing is wrong. A burned exhaust valve or certain cylinder head cracks would cause very low compression in that particular cylinder.
Answer B is wrong because while performing a compression test, a gradual buildup of compression with each stroke is normal and nothing is wrong. A burned exhaust valve or certain cylinder head cracks would cause very low compression in that particular cylinder.
Answer C is wrong because both technicians are wrong.
Answer D is correct because neither technician is right.

Question #91
Answer A is wrong because a burned exhaust valve would be a cause, not an indication.
Answer B is wrong because a cylinder with zero compression would not cause excessive exhaust noise.
Answer C is correct because excessive blowby could indicate a bad cylinder.
Answer D is wrong because backfire through the exhaust indicates incorrect ignition timing.

Question #92
Answer A is wrong because most leakage testers have two gauges: one for inlet air pressure and the other for cylinder pressure.
Answer B is correct because the second gauge indicates the air pressure in the cylinder.
Answer C is wrong because only one technician is right.
Answer D is wrong because one technician is right.

Question #93
Answer A is wrong because air escaping from the tailpipe would indicate a faulty exhaust valve.
Answer B is wrong because air escaping from the PCV system would indicate worn piston rings.
Answer C is correct because air escaping from the throttle body opening or from the carburetor indicates a leaking intake valve.
Answer D is wrong because bubbles would indicate a faulty head gasket.

Question #94
Answer A is a good choice because primary circuit test functions on an engine analyzer include coil input voltage, coil primary resistance, curb idle, and idle vacuum. Yet, it is wrong because both technicians are right.
Answer B is a good choice because the kV test measures the voltage required to jump the spark plug air gap. Yet, it is wrong because both technicians are right.
Answer C is correct because both technicians are right.
Answer D is wrong because both technicians are right.

Question #95
Answer A is correct because the snap test is used to check the voltage increase for each cylinder under load.
Answer B is wrong because the spark plug gap has to be checked with a feeler gauge.
Answer C is wrong because snap tests are only used on the secondary ignition circuit.
Answer D is wrong because this is not the reason a snap test is performed.

Question #96
Answer A is wrong because this is a normal condition and it should be looked into further.
Answer B is correct because if the carbon is excessive, it should be cleaned.
Answer C is wrong because only one technician is right.
Answer D is wrong because one technician is right.

Question #97
Answer A is wrong because with the rising cost of labor, it can be much cheaper to replace the cylinder head.
Answer B is correct because welding cylinder head cracks can save the customer the cost of a new or remanufactured one.
Answer C is wrong because only one technician is right.
Answer D is wrong because one technician is right.

Question #98
Answer A is wrong because the spring tension should be within 10 percent of the manufacturer's specification.
Answer B is correct because there should be no more than 10 pounds difference between the springs.
Answer C is wrong because only one technician is right.
Answer D is wrong because one technician is right.

Question #99
Answer A is wrong because the valve stem is unlikely to bend.
Answer B is wrong because the valve spring would not crack if the spring was not square.
Answer C is correct because the valve guide will be side loaded and cause premature wear.
Answer D is wrong because the free length of the spring would not be affected.

Question #100
Answer A is wrong because you do not inspect the valve lock grooves for carbon deposits.
Answer B is wrong because you do not inspect the valve lock grooves for groove thickness.
Answer C is wrong because you do not inspect the valve lock grooves for flat spots on the valve face.
Answer D is correct because you do inspect the valve lock grooves for uneven shoulders.

Question #101
Answer A is wrong because the valve is not the component being measured.
Answer B is correct because valve guide wear is being measured.
Answer C is wrong because when measuring the valve seat concentricity, the dial indicator must be on the valve seat.
Answer D is wrong because valve seat run out is the same as valve seat concentricity.

Question #102
Answer A is wrong because if pieces of the stone are missing, replace the stone.
Answer B is wrong because the stone should be dressed after 8 to 10 valves.
Answer C is wrong because the valve has to be cooled by the oil.
Answer D is correct because the grinding stone needs redressing if the valve chatters during the resurfacing process.

Question #103
Answer A is wrong because you do lift the grinding stone on and off at the rate of 120 times per minute.
Answer B is wrong because you do wait until the cutting stone starts to score the valve seat before resurfacing the stone.
Answer C is wrong because you only remove enough material to provide a new surface.
Answer D is correct because you never apply pressure to the grinding stone.

Question #104
Answer A is a good choice because you do measure the valve seat width with a machinist's ruler. Yet, it is wrong because both technicians are right.
Answer B is a good choice because you remove material according to the valve seat width measurement. Yet, it is wrong because both technicians are right.
Answer C is correct because both technicians are right.
Answer D is wrong because both technicians are right.

Question #105
Answer A is wrong because you use only emery paper to polish the valve stems; 240 grit is too rough.
Answer B is correct because you lubricate the valve stems with assembly lubricant, or at least with clean engine oil.
Answer C is wrong because only one technician is right.
Answer D is wrong because one technician is right.

Question #106
Answer A is correct and the exception, because cast iron is not a good material for a rocker arm due to its brittleness.
Answer B is wrong because stamped steel is also a good material for a rocker arm.
Answer C is wrong because titanium is also a good material for a rocker arm.
Answer D is wrong because aluminum is also a good material for a rocker arm.

Question #107
Answer A is a good choice because wear in that location indicates there was a lack of lubrication. Yet, it is wrong because both technicians are right.
Answer B is a good choice because the shaft itself cannot be checked for straightness. Yet, it is wrong because both technicians are right.
Answer C is correct because both technicians are right.
Answer D is wrong because both technicians are right.

Question #108
Answer A is wrong because the lifter is a solid roller type.
Answer B is correct because the rocker arm shaft should be checked for straightness.
Answer C is wrong because only one technician is right.
Answer D is wrong because one technician is right.

Question #109
Answer A is wrong because the temperature of the engine oil is not a concern during break-in.
Answer B is wrong because the thermostat opening has nothing to do with camshaft break-in.
Answer C is correct because enough oil is cooling and lubricating the contact surfaces of the lifter.
Answer D is wrong because the break-in procedure cannot be rushed.

Question #110
Answer A is wrong because the adjusting nut retains the rocker arm to a rocker arm stud.
Answer B is wrong because the figure shows an adjustable screw that follows the camshaft.
Answer C is wrong because adjustable pushrods are not available as yet.
Answer D is correct because an adjustable screw in the end of the rocker arm is not used.

Question #111
Answer A is correct because stamped steel gasket surfaces can be straightened.
Answer B is wrong because you never use anything but gasket sealer on engine gaskets.
Answer C is wrong because only one technician is right.
Answer D is wrong because one technician is right.

Question #112
Answer A is a good choice because when replacing a ECM/PCM, a ground strap and conductive mat must be used to ensure that there is no static electricity damage to the vehicle. Yet, it is wrong because both technicians are right.
Answer B is a good choice because you disconnect the vehicle's battery, and touching bare metal on the vehicle prior to replacement will protect the vehicle from static electricity damage. Yet, it is wrong because both technicians are right.
Answer C is correct because both technicians are right.
Answer D is wrong because both technicians are right.

Question #113
Answer A is wrong because a ground strap and conductive mat is not used to perform battery service.
Answer B is wrong because you do not use a ground strap and conductive mat when working on the secondary ignition system.
Answer C is wrong because when replacing the ignition coil, you do not use a ground strap and conductive mat.
Answer D is correct because when working with electronic control modules, you do use a grounding strap and conductive mat.

Glossary

ABS An abbreviation for Anti-lock Brake System.

Absolute Pressure The aero point from which pressure is measured.

Ackerman Principle The geometric principle used to provide toe-out on turns. The ends of the steering arms are angled so that the inside wheel turns more than the outside wheel when a vehicle is making a turn.

Actuator A device that delivers motion in response to an electrical signal.

Adapter The welds under a spring seat to increase the mounting height or fit a seal to the axle.

Adapter Ring A part that is bolted between the clutch cover and the flywheel on some two-plate clutches when the clutch is installed on a flat flywheel.

A/D Converter An abbreviation for Analog-to-Digital Converter.

Additive An additive intended to improve a certain characteristic of the material.

Adjustable Torque Arm A member used to retain axle alignment and, in some cases, control axle torque. Normally one adjustable and one rigid torque arm are used per axle so the axle can be aligned. This rod has means by which it can be extended or retracted for adjustment purposes.

Adjusting Ring A device that is held in the shift signal valve bore by a press fit pin through the valve body housing. When the ring is pushed in by the adjusting tool, the slots on the ring that engage the pin are released.

After-Cooler A device that removes water and oil from the air by a cooling process. The air leaving an after-cooler is saturated with water vapor, which condenses when a drop in temperature occurs.

Air Bag An air-filled device that functions as the spring on axles that utilize air pressure in the suspension system.

Air Brakes A braking system that uses air pressure to actuate the brakes by means of diaphragms, wedges, or cams.

Air Brake System A system utilizing compressed air to activate the brakes.

Air Compressor (1) An engine-driven mechanism for supplying high pressure air to the truck brake system. There are basically two types of compressors: those designed to work on in-line engines and those that work on V-type engines. The in-line type is mounted forward and is gear driven, while the V-type is mounted toward the firewall and is camshaft driven. With both types the coolant and lubricant are supplied by the truck engine. (2) A pump-like device in the air conditioning system that compresses refrigerant vapor to achieve a change in state for the refrigeration process.

Air Conditioning The control of air movement, humidity, and temperature by mechanical means.

Air Dryer A unit that removes moisture.

Air Filter/Regulator Assembly A device that minimizes the possibility of moisture-laden air or impurities entering a system.

Air Hose An air line, such as one between the tractor and trailer, that supplies air for the trailer brakes.

Air-Over-Hydraulic Brakes A brake system utilizing a hydraulic system assisted by an air pressure system.

Air-Over-Hydraulic Intensifier A device that changes the pneumatic air pressure from the treadle brake valve into hydraulic pressure which controls the wheel cylinders.

Air Shifting The process that uses air pressure to engage different range combinations in the transmission's auxiliary section without a mechanical linkage to the driver.

Air Slide Release A release mechanism for a sliding fifth wheel, which is operated from the cab of a tractor by actuating an air control valve. When actuated, the valve energizes an air cylinder, which releases the slide lock and permits positioning of the fifth wheel.

Air Spring An airfilled device that functions as the spring on axles that utilize air pressure in the suspension system.

Air Spring Suspension A single or multi-axle suspension relying on air bags for springs and weight distribution of axles.

Air Timing The time required for the air to be transmitted to or released from each brake, starting the instant the driver moves the brake pedal.

Altitude Compensation System An altitude barometric switch and solenoid used to provide better driveability at more than 4,000 feet (1220 meters) above sea level.

Ambient Temperature Temperature of the surrounding or prevailing air. Normally, it is considered to be the temperature in the service area where testing is taking place.

Amboid Gear A gear that is similar to the hypoid type with one exception: the axis of the drive pinion gear is located above the centerline axis of the ring gear.

Amp An abbreviation for ampere.

Ampere The unit for measuring electrical current.

Analog Signal A voltage signal that varies within a given range (from high to low, including all points in between).

Analog-to-Digital Converter (A/D converter) A device that converts analog voltage signals to a digital format; this is located in a section of the processor called the input signal conditioner.

Analog Volt/Ohmmeter (AVOM) A test meter used for checking voltage and resistance. Analog meters should not be used on solid state circuits.

Annulus The largest part of a simple gear set.

Anticorrosion Agent A chemical used to protect metal surfaces from corrosion.

Antifreeze A compound, such as alcohol or glycerin, that is added to water to lower its freezing point.

Anti-lock Brake System (ABS) A computer controlled brake system having a series of sensing devices at each wheel that control braking action to prevent wheel lockup.

Anti-lock Relay Valve (ARV) In an anti-lock brake system, the device that usually replaces the standard relay valve used to control the rear axle service brakes and performs the standard relay function during tractor/trailer operation.

Antirattle Springs Springs that reduce wear between the intermediate plate and the drive pin, and helps to improve clutch release.

Antirust Agent An additive used with lubricating oils to prevent rusting of metal parts when the engine is not in use.

Application Valve A foot-operated brake valve that controls air pressure to the service chambers.

Applied Moment A term meaning a given load has been placed on a frame at a particular point.

Area The total cross section of a frame rail including all applicable elements usually given in square inches.

Armature The rotating component of a (1) starter or other motor. (2) generator. (3) compressor clutch.

Articulating Upper Coupler A bolster plate kingpin arrangement that is not rigidly attached to the trailer, but provides articulation and/or oscillation, (such as a frameless dump) about an axis parallel to the rear axle of the trailer.

Articulation Vertical movement of the front driving or rear axle relative to the frame of the vehicle to which they are attached.

ASE An abbreviation for Automotive Service Excellence, a trademark of National Institute for Automotive Service Excellence.

Aspect Ratio A tire term calculated by dividing the tire's section height by its section width.

ATEC System A system that includes an electronic control system, torque converter, lockup clutch, and planetary gear train.

Atmospheric Pressure The weight of the air at sea level; 14.696 pounds per square inch (psi) or 101.33 kilopascals (kPa).

Automatic Slack Adjuster The device that automatically adjusts the clearance between the brake linings and the brake drum or rotor. The slack adjuster controls the clearance by sensing the length of the stroke of the push rod for the air brake chamber.

Autoshift Finger The device that engages the shift blocks on the yoke bars that corresponds to the tab on the end of the gearshift lever in manual systems.

Auxiliary Filter A device installed in the oil return line between the oil cooler and the transmission to prevent debris from being flushed into the transmission causing a failure. An auxiliary filter must be installed before the vehicle is placed back in service.

Auxiliary Section The section of a transmission where range shifting occurs, housing the auxiliary drive gear, auxiliary main shaft assembly, auxiliary countershaft, and the synchronizer assembly.

Axis of Rotation The center line around which a gear or part revolves.

Axle (1) A rod or bar on which wheels turn. (2) A shaft that transmits driving torque to the wheels.

Axle Range Interlock A feature designed to prevent axle shifting when the interaxle differential is locked out, or when lockout is engaged. The basic shift system operates the same as the standard shift system to shift the axle and engage or disengage the lockout.

Axle Seat A suspension component used to support and locate the spring on an axle.

Axle Shims Thin wedges that may be installed under the leaf springs of single axle vehicles to tilt the axle and correct the U-joint operating angles. Wedges are available in a range of sizes to change pinion angles.

Backing Plate A metal plate that serves as the foundation for the brake shoes and other drum brake hardware.

Battery Terminal A tapered post or threaded studs on top of the battery case, or infernally threaded provisions on the side of the battery for connecting the cables.

Beam Solid Mount Suspension A tandem suspension relying on a pivotal mounted beam, with axles attached at the ends for load equalization. The beam is mounted to a solid center pedestal.

Beam Suspension A tandem suspension relying on a pivotally mounted beam, with axles attached at ends for lead equalization. Beam is mounted to center spring.

Bellows A movable cover or seal that is pleated or folded like an accordion to allow for expansion and contraction.

Bending Moment A term implying that when a load is applied to the frame, it will be distributed across a given section of the frame material.

Bias A tire term where belts and plies are laid diagonally or crisscrossing each other.

Bimetallic Two dissimilar metals joined together that have different bending characteristics when subjected to different changes of temperature.

Blade Fuse A type of fuse having two flat male lugs sticking out for insertion in the female box connectors.

Bleed Air Tanks The process of draining condensation from air tanks to increase air capacity and brake efficiency.

Block Diagnosis Chart A troubleshooting chart that lists symptoms, possible causes, and probable remedies in columns.

Blower Fan A fan that pushes or blows air through a ventilation, heater, or air conditioning system.

Bobtail Proportioning Valve A valve that senses when the tractor is bobtailing and automatically reduces the amount of air pressure that can be applied to the tractor's drive axle(s). This reduces braking force on the drive axles, lessening the chance of a spin out on slippery pavement.

Bobtailing A tractor running without a trailer.

Bogie The axle spring, suspension arrangement on the rear of a tandem axle tractor.

Bolster Plate The flat load-bearing surface under the front of a semitrailer, including the kingpin, which rests firmly on the fifth wheel when coupled.

Bolster Plate Height The height from the ground to the bolster plate when the trailer is level and empty.

Boss A heavy cast section that is used for support, such as the outer race of a bearing.

Bottoming A condition that occurs when; (1) The teeth of one gear touch the lowest point between teeth of a mating gear. (2) The bed or frame of the vehicle strikes the axle, such as may be the case of overloading.

Bottom U-Bolt Plate A plate that is located on the bottom side of the spring or axle and is held in place when the U-bolts are tightened to the clamp spring and axle together.

Bracket An attachment used to secure parts to the body or frame.

Brake Control Valve A dual brake valve that releases air from the service reservoirs to the service lines and brake chambers. The

Appendices
Glossary

valve includes a piston which pushes on diaphragms to open ports; these vent air to service lines in the primary and secondary systems.

Brake Disc A steel disc used in a braking system with a caliper and pads. When the brakes are applied, the pad on each side of the spinning disc is forced against the disc, thus imparting a braking force. This type of brake is very resistant to brake fade.

Brake Drum A cast metal bell-like cylinder attached to the wheel that is used to house the brake shoes and provide a friction surface for stopping a vehicle.

Brake Fade A condition that occurs when friction surfaces become hot enough to cause the coefficient of friction to drop to a point where the application of severe pedal pressure results in little actual braking.

Brake Lining A special friction material used to line brake shoes or brake pads. It withstands high temperatures and pressure. The molded material is either riveted or bonded to the brake shoe, with a suitable coefficient of friction for stopping a vehicle.

Brake Pad The friction lining and plate assembly that is forced against the rotor to cause braking action in a disc brake system.

Brake Shoe The curved metal part, faced with brake lining, which is forced against the brake drum to produce braking action.

Brake Shoe Rollers A hardware part that attaches to the web of the brake shoes by means of roller retainers. The rollers, in turn, ride on the end of an S-cam.

Brake System The vehicle system that slows or stops a vehicle. A combination of brakes and a control system.

Breakaway Valve A device that automatically seals off the tractor air supply from the trailer air supply when the tractor system pressure drops to 30 or 40 psi (207 to 276 kPa).

British Thermal Unit (Btu) A measure of heat quantity equal to the amount of heat required to raise 1 pound of water 1°F.

Broken Back Drive Shaft A term often used for non-parallel drive shaft.

Btu An abbreviation for British Thermal Unit.

Bump Steer Erratic steering caused from rolling over bumps, cornering, or heavy braking. Same as orbital steer and roll steer.

CAA An abbreviation for Clean Air Act.

Caliper A disc brake component that changes hydraulic pressure into mechanical force and uses that force to press the brake pads against the rotor and stop the vehicle. Calipers come in three basic types: fixed, floating, and sliding, and can have one or more pistons.

Camber The attitude of a wheel and tire assembly when viewed from the front of a car. If it leans outward, away from the car at the top, the wheel is said to have positive camber. If it leans inward, it is said to have negative camber.

Cam Brakes Brakes that are similar in operation and design to the wedge brake, with the exception that an S-type camshaft is used instead of a wedge and rubber assembly.

Cartridge Fuse A type of fuse having a strip of low melting point metal enclosed in a glass tube. If an excessive current flows through the circuit, the fuse element melts at the narrow portion, opening the circuit and preventing damage.

Caster The angle formed between the kingpin axis and a vertical axis as viewed from the side of the vehicle. Caster is considered positive when the top of the kingpin axis is behind the vertical axis.

Cavitation A condition that causes bubble formation.

Center of Gravity The point around which the weight of a truck is evenly distributed; the point of balance.

Ceramic Fuse A fuse found in some import vehicles that has a ceramic insulator with a conductive metal strip along one side.

CFC An abbreviation for chlorofluorocarbon.

Charging System A system consisting of the battery, alternator, voltage regulator, associated wiring, and the electrical loads of a vehicle. The purpose of the system is to recharge the battery whenever necessary and to provide the current required to power the electrical components.

Charge the Trailer To supply the trailer air tanks with air by means of a dash control valve, tractor protection valve, and a trailer relay emergency valve.

Charging Circuit The alternator (or generator) and associated circuit used to keep the battery charged and to furnish power to the vehicle's electrical systems when the engine is running.

Check Valve A valve that allows air to flow in one direction only. It is a federal requirement to have a check valve between the wet and dry air tanks.

Chlorofluorocarbon (CFC) A compound used in the production of refrigerant that is believed to cause damage to the ozone layer.

Circuit The complete path of an electrical current, including the generating device. When the path is unbroken, the circuit is closed and current flows. When the circuit continuity is broken, the circuit is open and current flow stops.

Clean Air Act (CAA) Federal regulations, passed in 1992, that have resulted in major changes in air-conditioning systems.

Climbing A gear problem caused by excessive wear in gears, bearings, and shafts whereby the gears move sufficiently apart to cause the apex (or point) of the teeth on one gear to climb over the apex of the teeth on another gear with which it is meshed.

Clutch A device for connecting and disconnecting the engine from the transmission or for a similar purpose in other units.

Clutch Brake A circular disc with a friction surface that is mounted on the transmission input spline between the release bearing and the transmission. Its purpose is to slow or stop the transmission input shaft from rotating in order to allow gears to be engaged without clashing or grinding.

Clutch Housing A component that surrounds and protects the clutch and connects the transmission case to the vehicle's engine.

Clutch Pack An assembly of normal clutch plates, friction discs, and one very thick plate known as the pressure plate. The pressure plate has tabs around the outside diameter to mate with the channel in the clutch drum.

COE An abbreviation for cab-over-engine.

Coefficient of Friction A measurement of the amount of friction developed between two objects in physical contact when one of the objects is drawn across the other.

Coil Springs Spring steel spirals that are mounted on control arms or axles to absorb road shock.

Combination A truck coupled to one or more trailers.

Compression Applying pressure to a spring or any springy substance, thus causing it to reduce its length in the direction of the compressing force.

Compressor (1) A mechanical device that increases pressure within a container by pumping air into it. (2) That component of an air-conditioning system that compresses low temperature/pressure refrigerant vapor.

Condensation The process by which gas (or vapor) changes to a liquid.

Condenser A component in an air-conditioning system used to cool a refrigerant below its boiling point causing it to change from a vapor to a liquid.

Conductor Any material that permits the electrical current to flow.

Constant Rate Springs Leaf-type spring assemblies that have a constant rate of deflection.

Control Arm The main link between the vehicle's frame and the wheels that acts as a hinge to allow wheel action up and down independent of the chassis.

Controlled Traction A type of differential that uses a friction plate assembly to transfer drive torque from the vehicle's slipping wheel to the one wheel that has good traction or surface bite.

Converter Dolly An axle, frame, drawbar, and fifth wheel arrangement that converts a semitrailer into a full trailer.

Coolant Liquid that circulates in an engine cooling system.

Coolant Heater A component used to aid engine starting and reduce the wear caused by cold starting.

Coolant Hydrometer A tester designed to measure coolant specific gravity and determine the amount of antifreeze in the coolant.

Cooling System Complete system for circulating coolant.

Coupling Point The point at which the turbine is turning at the same speed as the impeller.

Crankcase The housing within which the crankshaft and many other parts of the engine operate.

Cranking Circuit The starter and its associated circuit, including battery, relay (solenoid), ignition switch, neutral start switch (on vehicles with automatic transmission), and cables and wires.

Cross Groove Joint disc-shaped type of inner CV joint that uses balls and V-shaped grooves on the inner and outer races to accommodate the plunging motion of the half-shaft. The joint usually bolts to a transaxle stub flange; same as disc-type joint.

Cross-Tube A system that transfers the steering motion to the opposite, passenger side steering knuckle. It links the two steering knuckles together and forces them to act in unison.

C-Train A combination of two or more trailers in which the dolly is connected to the trailer by means of two pintle hook or coupler drawbar connections. The resulting connection has one pivot point.

Cycling (1) Repeated on-off action of the air conditioner compressor. (2) Heavy and repeated electrical cycling that can cause the positive plate material to break away from its grids and fall into the sediment chambers at the base of the battery case.

Dampen To slow or reduce oscillations or movement.

Dampened Discs Discs that have dampening springs incorporated into the disc hub. When engine torque is first transmitted to the disc, the plate rotates on the hub, compressing the springs. This action absorbs the shocks and torsional vibration caused by today's low rpm, high torque, engines.

Dash Control Valves A variety of handoperated valves located on the dash. They include parking brake valves, tractor protection valves, and differential lock.

Data Links Circuits through which computers communicate with other electronic devices such as control panels, modules, some sensors, or other computers in the form of digital signals.

Dead Axle Non-live or dead axles are often mounted in lifting suspensions. They hold the axle off the road when the vehicle is traveling empty, and put it on the road when a load is being carried. They are also used as air suspension third axles on heavy straight trucks and are used extensively in eastern states with high axle weight laws. An axle that does not rotate but merely forms a base on which to attach the wheels.

Deadline To take a vehicle out of service.

Deburring To remove sharp edges from a cut.

Dedicated Contract Carriage Trucking operations set up and run according to a specific shipper's needs. In addition to transportation, they often provide other services such as warehousing and logistics planning.

Deflection Bending or moving to a new position as the result of an external force.

Department of Transportation (DOT) A government agency that establishes vehicle standards.

Detergent Additive An additive that helps keep metal surfaces clean and prevents deposits. These additives suspend particles of carbon and oxidized oil in the oil.

DER An abbreviation for Department of Environmental Resources.

Diagnostic Flow Chart A chart that provides a systematic approach to the electrical system and component troubleshooting and repair. They are found in service manuals and are vehicle make and model specific.

Dial Caliper A measuring instrument capable of taking inside, outside, depth, and step measurements.

Differential A gear assembly that transmits power from the drive shaft to the wheels and allows two opposite wheels to turn at different speeds for cornering and traction.

Differential Carrier Assembly An assembly that controls the drive axle operation.

Differential Lock A toggle or push-pull type air switch that locks together the rear axles of a tractor so they pull as one for off-the-road operation.

Digital Binary Signal A signal that has only two values; on and off.

Digital Volt/Ohmmeter (DVOM) A type of test meter recommended by most manufacturers for use on solid state circuits.

Diode The simplest semiconductor device formed by joining P-type semiconductor material with N-type semiconductor material. A diode allows current to flow in one direction, but not in the opposite direction.

Direct Drive The gearing of a transmission so that in its highest gear, one revolution of the engine produces one revolution of the transmission's output shaft. The top gear or final drive ratio of a direct drive transmission would be 1:1.

Disc Brake A steel disc used in a braking system with a caliper and pads. When the brakes are applied, the pad on each side of the spinning disc is forced against the disc, thus imparting a braking

Dispatch Sheet A form used to keep track of dates when the work is to be completed. Some dispatch sheets follow the job through each step of the servicing process.

Dog Tracking Off-center tracking of the rear wheels as related to the front wheels.

DOT An abbreviation for Department of Transportation.

Downshift Control The selection of a lower range to match driving conditions encountered or expected to be encountered. Learning to take advantage of a downshift gives better control on slick or icy roads and on steep downgrades. Downshifting to lower ranges increases engine braking.

Double Reduction Axle An axle that uses two gear sets for greater overall gear reduction and peak torque development. This design is favored for severe service applications, such as dump trucks, cement mixers, and other heavy haulers.

Drag Link A connecting rod or link between the steering gear, Pitman arm, and the steering linkage.

Drawbar Capacity The maximum, horizontal pulling force that can be safely applied to a coupling device.

Driven Gear A gear that is driven or forced to turn by a drive gear, by a shaft, or by some other device.

Drive or Driving Gear A gear that drives another gear or causes another gear to turn.

Drive Line The propeller or drive shaft, universal joints, and so forth, that links the transmission output to the axle pinion gear shaft.

Drive Line Angle The alignment of the transmission output shaft, driveshaft, and rear axle pinion centerline.

Drive Shaft An assembly of one or two universal joints connected to a shaft or tube; used to transmit power from the transmission to the differential.

Drive Train An assembly that includes all power transmitting components from the rear of the engine to the wheels, including clutch/torque converter, transmission, drive line, and front and rear driving axles.

Driver Controlled Main Differential Lock A type of axle assembly has greater flexibility over the standard type of single reduction axle because it provides equal amounts of drive line torque to each driving wheel, regardless of changing road conditions. This design also provides the necessary differential action to the road wheels when the truck is turning a corner.

Driver's Manual A publication that contains information needed by the driver to understand, operate, and care for the vehicle and its components.

Drum Brake A type of brake system in which stopping friction is created by the shoes pressing against the inside of the rotating drum.

Dual Hydraulic Braking System A brake system consisting of a tandem, or double action master cylinder which is basically two master cylinders usually formed by aligning two separate pistons and fluid reservoirs into a single cylinder.

ECU An abbreviation for electronic control unit.

Eddy Current A small circular current produced inside a metal core in the armature of a starter motor. Eddy currents produce heat and are reduced by using a laminated core.

Electricity The movement of electrons from one place to another.

Electric Retarder Electromagnets mounted in a steel frame. Energizing the retarder causes the electromagnets to exert a dragging force on the rotors in the frame and this drag force is transmitted directly to the drive shaft.

Electromotive Force (EMF) The force that moves electrons between atoms. This force is the pressure that exists between the positive and negative points (the electrical imbalance). This force is measured in units called volts.

Electronically Programmable Memory (EPROM) Computer memory that permits adaptation of the ECU to various standard mechanically controlled functions.

Electronic Control Unit (ECU) The brain of the vehicle.

Electronics The technology of controlling electricity.

Electrons Negatively charged particles orbiting around every nucleus.

Elliot Axle A solid bar front axle on which the ends span the steering knuckle.

EMF An abbreviation for electromotive force.

End Yoke The component connected to the output shaft of the transmission to transfer engine torque to the drive shaft.

Engine Brake A hydraulically operated device that converts the vehicle's engine into a power absorbing retarding mechanism.

Engine Stall Point The point, in rpms, under load is compared to the engine manufacturer's specified rpm for the stall test.

Environmental Protection Agency An agency of the United States government charged with the responsibilities of protecting the environment and enforcing the Clean Air Act (CAA) of 1990.

EPA An abbreviation for the Environmental Protection Agency.

EPROM An abbreviation for Electronically Programmable Memory.

Equalizer A suspension device used to transfer and maintain equal load distribution between two or more axles of a suspension. Formerly called a rocker beam.

Equalizer Bracket A bracket for mounting the equalizer beam of a multiple axle spring suspension to a truck or trailer frame while allowing for the beam's pivotal movement. Normally there are three basic types: flange-mount, straddle-mount, and under- or side-mount.

Evaporator A component in an air conditioning system used to remove heat from the air passing through it.

Exhaust Brake A valve in the exhaust pipe between the manifold and the muffler. A slide mechanism which restricts the exhaust flow, causing exhaust back pressure to build up in the engine's cylinders. The exhaust brake actually transforms the engine into a low pressure air compressor driven by the wheels.

External Housing Damper A counterweight attached to an arm on the rear of the transmission extension housing and designed to dampen unwanted driveline or powertrain vibrations.

Extra Capacity A term that generally refers to: (1) A coupling device that has strength capability greater than standard. (2) An oversized tank or reservoir for a fluid or vapor.

False Brinelling The polishing of a surface that is not damaged.

Fanning the Brakes Applying and releasing the brakes in rapid succession on a long downgrade.

Fatigue Failures The progressive destruction of a shaft or gear teeth material usually caused by overloading.

Fault Code A code that is recorded into the computer's memory. A fault code can be read by plugging a special break-out box tester into the computer.

Federal Motor Vehicle Safety Standard (FMVSS) A federal standard that specifies that all vehicles in the United States be assigned a Vehicle Identification Number (VIN).

Federal Motor Vehicle Safety Standard No. 121 (FMVSS 121) A federal standard that made significant changes in the guidelines that cover air brake systems. Generally speaking, the requirements of FMVSS 121 are such that larger capacity brakes and heavier steerable axles are needed to meet them.

FHWA An abbreviation for Federal Highway Administration.

Fiber Composite Springs Springs that are made of fiberglass, laminated, and bonded together by tough polyester resins.

Fifth Wheel A coupling device mounted on a truck and used to connect a semitrailer. It acts as a hinge point to allow changes in direction of travel between the tractor and the semitrailer.

Fifth Wheel Height The distance from the ground to the top of the fifth wheel when it is level and parallel with the ground. It can also refer to the height from the tractor frame to the top of the fifth wheel. The latter definition applies to data given in fifth wheel literature.

Fifth Wheel Top Plate The portion of the fifth wheel assembly that contacts the trailer bolster plate and houses the locking mechanism that connects to the kingpin.

Final Drive The last reduction gear set of a truck.

Fixed Value Resistor An electrical device that is designed to have only one resistance rating, which should not change, for controlling voltage.

Flammable Any material that will easily catch fire or explode.

Flare To spread gradually outward in a bell shape.

Flex Disc A term often used for flex plate.

Flex Plate A component used to mount the torque converter to the crankshaft. The flex plate is positioned between the engine crankshaft and the T/C. The purpose of the flex plate is to transfer crankshaft rotation to the shell of the torque converter assembly.

Float A cruising drive mode in which the throttle setting matches engine speed to road speed, neither accelerating nor decelerating.

Floating Main Shaft The main shaft consisting of a heavy-duty central shaft and several gears that turn freely when not engaged. The main shaft can move to allow for equalization of the loading on the countershafts. This is key to making a twin countershaft transmission workable. When engaged, the floating main shaft transfers torque evenly through its gears to the rest of the transmission and ultimately to the rear axle.

FMVSS An abbreviation for Federal Motor Vehicle Safety Standard.

FMVSS No 121 An abbreviation for Federal Motor Vehicle Safety Standard No 121.

Foot Valve A foot-operated brake valve that controls air pressure to the service chambers.

Foot-Pound An English unit of measurement for torque. One foot-pound is the torque obtained by a force of 1 pound applied to a foot long wrench handle.

Forged Journal Cross Part of a universal joint.

Frame Width The measurement across the outside of the frame rails of a tractor, truck, or trailer.

Franchised Dealership A dealer that has signed a contract with a particular manufacturer to sell and service a particular line of vehicles.

Fretting A result of vibration that the bearing outer race can pick up the machining pattern.

Friction Plate Assembly An assembly consisting of a multiple disc clutch that is designed to slip when a predetermined torque value is reached.

Front Axle Limiting Valve A valve that reduces pressure to the front service chambers, thus eliminating front wheel lockup on wet or icy pavements.

Front Hanger A bracket for mounting the front of the truck or trailer suspensions to the frame. Made to accommodate the end of the spring on spring suspensions. There are four basic types: flange-mount, straddle-mount, under-mount, and side-mount.

Full Trailer A trailer that does not transfer load to the towing vehicle. It employs a tow bar coupled to a swiveling or steerable running gear assembly at the front of the trailer.

Fully Floating Axles An axle configuration whereby the axle half shafts transmit only driving torque to the wheels and not bending and torsional loads that are characteristic of the semi-floating axle.

Fully Oscillating Fifth Wheel A fifth wheel type with fore/aft and side-to-side articulation.

Fusible Link A term often used for fuse link.

Fuse Link A short length of smaller gauge wire installed in a conductor, usually close to the power source.

GCW An abbreviation for gross combination weight.

Gear A disk-like wheel with external or internal teeth that serves to transmit or change motion.

Gear Pitch The number of teeth per given unit of pitch diameter, an important factor in gear design and operation.

General Over-the-Road Use A fifth wheel designed for multiple standard duty highway applications.

Gladhand The connectors between tractor and trailer air lines.

Gross Combination Weight (GCW) The total weight of a fully quipped vehicle including payload, fuel, and driver.

Gross Trailer Weight (GTW) The sum of the weight of an empty trailer and its payload.

Gross Vehicle Weight (GVW) The total weight of a fully equipped vehicle and its payload.

Ground The negatively charged side of a circuit. A ground can be a wire, the negative side of the battery, or the vehicle chassis.

Grounded Circuit A shorted circuit that causes current to return to the battery before it has reached its intended destination.

GTW An abbreviation for gross trailer weight.

GVW An abbreviation for gross vehicle weight.

Halogen Light A lamp having a small quartz/glass bulb that contains a fuel filament surrounded by halogen gas. It is contained within a larger metal reflector and lens element.

Hand Valve (1) A valve mounted on the steering column or dash, used by the driver to apply the trailer brakes independently of the tractor brakes. (2) A hand operated valve used to control the flow of fluid or vapor.

Harness and Harness Connectors The organization of the vehicle's electrical system providing an orderly and convenient starting point for tracking and testing circuits.

Hazardous Materials Any substance that is flammable, explosive, or is known to produce adverse health effects in people or the environment that are exposed to the material during its use.

Heads Up Display (HUD) A technology used in some vehicles that superimposes data on the driver's normal field of vision. The operator can view the information, which appears to "float" just above the hood at a range near the front of a conventional tractor or truck. This allows the driver to monitor conditions such as limited road speed without interrupting his normal view of traffic.

Heater Control Valve A valve that controls the flow of coolant into the heater core from the engine.

Heat Exchanger A device used to transfer heat, such as a radiator or condenser.

Heavy-Duty Truck A truck that has a GVW of 26,001 pounds or more.

Helper Spring An additional spring device that permits greater load on an axle.

High CG Load Any application in which the load center of gravity (CG) of the trailer exceeds 40 inches (102 centimeters) above the top of the fifth wheel.

High-Resistant Circuits Circuits that have an increase in circuit resistance, with a corresponding decrease in current.

High-Strength Steel A low-alloy steel that is much stronger than hot-rolled or cold-rolled sheet steels that normally are used in the manufacture of car body frames.

Hinged Pawl Switch The simplest type of switch; one that makes or breaks the current of a single conductor.

HUD An abbreviation for heads up display.

Hydraulic Brakes Brakes that are actuated by a hydraulic system.

Hydraulic Brake System A system utilizing the properties of fluids under pressure to activate the brakes.

Hydrometer A tester designed to measure the specific gravity of a liquid.

Hypoid Gears Gears that intersect at right angles when meshed. Hypoid gearing uses a modified spiral bevel gear structure that allows several gear teeth to absorb the driving power and allows the gears to run quietly. A hypoid gear is typically found at the drive pinion gear and ring gear interface.

I-Beam Axle An axle designed to give great strength at reasonable weight. The cross section of the axle resembles the letter "I."

ICC Check Valve A valve that allows air to flow in one direction only. It is a federal requirement to have a check valve between the wet and dry air tanks.

Inboard Toward the centerline of the vehicle.

In-Line Fuse A fuse that is in series with the circuit in a small plastic fuse holder, not in the fuse box or panel. It is used, when necessary, as a protection device for a portion of the circuit even though the entire circuit may be protected by a fuse in the fuse box or panel.

In-Phase The in-line relationship between the forward coupling shaft yoke and the driveshaft slip yoke of a two-piece drive line.

Input Retarder A device located between the torque converter housing and the main housing designed primarily for over-the-road operations. The device employs a "paddle wheel" type design with a vaned rotor mounted between stator vanes in the retarder housing.

Installation Templates Drawings supplied by some vehicle manufacturers to allow the technician to correctly install the accessory. The templates available can be used to check clearances or to ease installation.

Insulator A material, such as rubber or glass, that offers high resistance to the flow of electrons.

Integrated Circuit A component containing diodes, transistors, resistors, capacitors, and other electronic components mounted on a single piece of material and capable to perform numerous functions.

Jacobs Engine Brake A term sometimes used for Jake brake.

Jake Brake The Jacobs engine brake, named for its inventor. A hydraulically operated device that converts a power producing diesel engine into a power-absorbing retarder mechanism by altering the engine's exhaust valve opening time used to slow the vehicle.

Jumper Wire A wire used to temporarily bypass a circuit or components for electrical testing. A jumper wire consists of a length of wire with an alligator clip at each end.

Jumpout A condition that occurs when a fully engaged gear and sliding clutch are forced out of engagement.

Jump Start The procedure used when it becomes necessary to use a booster battery to start a vehicle having a discharged battery.

Kinetic Energy Energy in motion.

Kingpin (1) The pin mounted through the center of the trailer upper coupler (bolster plate) that mates with the fifth wheel locks, securing the trailer to the fifth wheel. The configuration is controlled by industry standards. (2) A pin or shaft on which the steering spindle rotates.

Landing Gear The retractable supports for a semitrailer to keep the trailer level when the tractor is detached from it.

Lateral Runout The wobble or side-to-side movement of a rotating wheel or of a rotating wheel and tire assembly.

Lazer Beam Alignment System A two- or four-wheel alignment system using wheel-mounted instruments to project a lazer beam to measure toe, caster, and camber.

Lead The tendency of a car to deviate from a straight path on a level road when there is no pressure on the steering wheel in either direction.

Leaf Springs Strips of steel connected to the chassis and axle to isolate the vehicle from road shock.

Less Than Truckload (LTL) Partial loads from the networks of consolidation centers and satellite terminals.

Light Beam Alignment System An alignment system using wheel-mounted instruments to project light beams onto charts and scales to measure toe, caster, and camber, and note the results of alignment adjustments.

Limited-Slip Differential A differential that utilizes a clutch device to deliver power to either rear wheel when the opposite wheel is spinning.

Linkage A system of rods and levers used to transmit motion or force.

Live Axle An axle on which the wheels are firmly affixed. The axle drives the wheels.

Live Beam Axle A non-independent suspension in which the axle moves with the wheels.

Load Proportioning Valve (LPV) A valve used to redistribute hydraulic pressure to front and rear brakes based on vehicle loads. This is a load- or height-sensing valve that senses the vehicle load and proportions the braking between front and rear brakes in proportion to the load variations and degree of rear-to-front weight transfer during braking.

Lockstrap A manual adjustment mechanism that allows for the adjustment of free travel.

Lock-up Torque Converter A torque converter that eliminates the 10 percent slip that takes place between the impeller and turbine at the coupling stage of operation. It is considered a four-element (impeller, turbine, stator, lockup clutch), three-stage (stall, coupling, and locking stage) unit.

Longitudinal Leaf Spring A leaf spring that is mounted so it is parallel to the length of the vehicle.

Low-Maintenance Battery A conventionally vented, lead/acid battery, requiring normal periodic maintenance.

LTL An abbreviation for less than truckload.

Magnetorque An electromagnetic clutch.

Maintenance-Free Battery A battery that does not require the addition of water during normal service life.

Maintenance Manual A publication containing routine maintenance procedures and intervals for vehicle components and systems.

Main Transmission A transmission consisting of an input shaft, floating main shaft assembly and main drive gears, two counter shaft assemblies, and reverse idler gears.

Manual Slide Release The release mechanism for a sliding fifth wheel, which is operated by hand.

Metering Valve A valve used on vehicles equipped with front disc and rear drum brakes. It improves braking balance during light brake applications by preventing application of the front disc brakes until pressure is built up in the hydraulic system.

Moisture Ejector A valve mounted to the bottom or side of the supply and service reservoirs that collects water and expels it every time the air pressure fluctuates.

Mounting Bracket That portion of the fifth wheel assembly that connects the fifth wheel top plate to the tractor frame or fifth wheel mounting system.

Multiaxle Suspension A suspension consisting of more than three axles.

Multiple Disc Clutch A clutch having a large drum-shaped housing that can be either a separate casting or part of the existing transmission housing.

NATEF An abbreviation for National Automotive Education Foundation.

National Automotive Education Foundation (NATEF) A foundation having a program of certifying secondary and post secondary automotive and heavy-duty truck training programs.

National Institute for Automotive Service Excellence (ASE) A nonprofit organization that has an established certification program for automotive, heavy-duty truck, auto body repair, engine machine shop technicians, and parts specialists.

Needlenose Pliers This tool has long tapered jaws for grasping small parts or for reaching into tight spots. Many needlenose pliers also have cutting edges and a wire stripper.

NIASE An abbreviation for National Institute for Automotive Service Excellence, now abbreviated ASE.

NIOSH An abbreviation for National Institute for Occupation Safety and Health.

NLGI An abbreviation for National Lubricating Grease Institute.

NHTSA An abbreviation for National Highway Traffic Safety Administration.

Nonlive Axle Non-live or dead axles are often mounted in lifting suspensions. They hold the axle off the road when the vehicle is traveling empty, and put it on the road when a load is being carried. They are also used as air suspension third axles on heavy straight trucks and are used extensively in eastern states with high axle weight laws.

Nonparallel Driveshaft A type of drive shaft installation whereby the working angles of the joints of a given shaft are equal; however the companion flanges and/or yokes are not parallel.

Nonpolarized Gladhand A gladhand that can be connected to either service or emergency gladhand.

Nose The front of a semitrailer.

No-tilt Convertible Fifth A fifth wheel with fore/aft articulation that can be locked out to produce a rigid top plate for applications that have either rigid and/or articulating upper couplers.

(OEM) An abbreviation for original equipment manufacturer.

Off-road With reference to unpaved, rough, or ungraded terrain on which a vehicle will operate. Any terrain not considered part of the highway system falls into this category.

Ohm A unit of measured electrical resistance.

Ohm's Law The basic law of electricity stating that in any electrical circuit, current, resistance, and pressure work together in a mathematical relationship.

On-road With reference to paved or smooth-graded surface terrain on which a vehicle will operate, generally considered to be part of the public highway system.

Open Circuit An electrical circuit whose path has been interrupted or broken either accidentally (a broken wire) or intentionally (a switch turned off).

Operational Control Valve A valve used to control the flow of compressed air through the brake system.

Oscillation The rotational movement in either fore/aft or side-to-side direction about a pivot point. Generally refers to fifth wheel designs in which fore/aft and side-to-side articulation are provided.

OSHA An abbreviation for Occupational Safety and Health Administration.

Out-of-Phase A condition of the universal joint which acts somewhat like one person snapping a rope held by a person at the opposite end. The result is a violent reaction at the opposite end. If both were to snap the rope at the same time, the resulting waves cancel each other and neither would feel the reaction.

Out-of-Round A wheel or tire defect in which the wheel or tire is not round.

Output Driver An electronic on/off switch that the computer uses to control the ground circuit of a specific actuator. Output drivers are located in the processor along with the input conditioners, microprocessor, and memory.

Output Yoke The component that serves as a connecting link, transferring torque from the transmission's output shaft through the vehicle's drive line to the rear axle.

Oval A condition that occurs when a tube is not round, but is somewhat egg-shaped.

Overall Ratio The ratio of the lowest to the highest forward gear in the transmission.

Overdrive The gearing of a transmission so that in its highest gear one revolution of the engine produces more than one revolution of the transmission's output shaft.

Overrunning Clutch A clutch mechanism that transmits power in one direction only.

Overspeed Governor A governor that shuts off the fuel or stops the engine when excessive speed is reached.

Oxidation Inhibitor (1) An additive used with lubricating oils to keep oil from oxidizing even at very high temperatures. (2) An additive for gasoline to reduce the chemicals in gasoline that react with oxygen.

Pad A disc brake lining and metal back riveted, molded, or bonded together.

Parallel Circuit An electrical circuit that provides two or more paths for the current to flow. Each path has separate resistors and operates independently from the other parallel paths. In a parallel circuit, amperage can flow through more than one resistor at a time.

Parallel Joint Type A type of drive shaft installation whereby all companion flanges and/or yokes in the complete drive line are parallel to each other with the working angles of the joints of a given shaft being equal and opposite.

Parking Brake A mechanically applied brake used to prevent a parked vehicle's movement.

Parts Requisition A form that is used to order new parts, on which the technician writes the names of what part(s) are needed along with the vehicle's VIN or company's identification folder.

Payload The weight of the cargo carried by a truck, not including the weight of the body.

Pipe or Angle Brace Extrusions between opposite hangers on a spring or air-type suspension.

Pitman Arm A steering linkage component that connects the steering gear to the linkage at the left end of the center link.

Pitting Surface irregularities resulting from corrosion.

Planetary Drive A planetary gear reduction set where the sun gear is the drive and the planetary carrier is the output.

Planetary Gear Set A system of gearing that is somewhat like the solar system. A pinion is surrounded by an internal ring gear and planet gears are in mesh between the ring gear and pinion around which all revolve.

Planetary Pinion Gears Small gears fitted into a framework called the planetary carrier.

Plies The layers of rubber-impregnated fabric that make up the body of a tire.

Pogo Stick The air and electrical line support rod mounted behind the cab to keep the lines from dragging between the tractor and trailer.

Polarity The particular state, either positive or negative, with reference to the two poles or to electrification.

Pole The number of input circuits made by an electrical switch.

Pounds per Square Inch (psi) A unit of English measure for pressure.

Power A measure of work being done.

Power Flow The flow of power from the input shaft through one or more sets of gears, or through an automatic transmission to the output shaft.

Power Steering A steering system utilizing hydraulic pressure to reduce the turning effort required of the operator.

Power Synchronizer A device to speed up the rotation of the main section gearing for smoother automatic downshifts and to slow down the rotation of the main section gearing for smoother automatic upshifts.

Power Train An assembly consisting of a drive shaft, coupling, clutch, and transmission differential.

Pressure The amount of force applied to a definite area measured in pounds per square inch (psi) English or kilopascals (kPa) metric.

Pressure Differential The difference in pressure between any two points of a system or a component.

Pressure Relief Valve (1) A valve located on the wet tank, usually preset at 150 psi (1,034 kPa). Limits system pressure if the compressor or governor unloader valve malfunctions. (2) A valve located on the rear head of an air-conditioning compressor or pressure vessel that opens if an excessive system pressure is exceeded.

Printed Circuit Board An electronic circuit board made of thin nonconductive plastic-like material onto which conductive metal, such as copper, has been deposited. Parts of the metal are then etched away by an acid, leaving metal lines that form the conductors for the various circuits on the board. A printed circuit board can hold many complex circuits in a very small area.

Programmable Read Only Memory (PROM) An electronic component that contains program information specific to different vehicle model calibrations.

PROM An abbreviation for Programmable Read Only Memory.

Priority Valve A valve that ensures that the control system upstream from the valve will have sufficient pressure during shifts to perform its automatic functions.

Proportioning Valve A valve used on vehicles equipped with front disc and rear drum brakes. It is installed in the lines to the rear drum brakes, and in a split system, below the pressure differential valve. By reducing pressure to the rear drum brakes, the valve helps to prevent premature lockup during severe brake application and provides better braking balance.

Psi An abbreviation for pounds per square inch.

Pull Circuit A circuit that brings the cab from a fully tilted position up and over the center.

Pull-Type Clutch A type of clutch that does not push the release bearing toward the engine; instead, it pulls the release bearing toward the transmission.

Pump/Impeller Assembly The input (drive) member that receives power from the engine.

Push Circuit A circuit that raises the cab from the lowered position to the desired tilt position.

Push-Type Clutch A type of clutch in which the release bearing is not attached to the clutch cover.

P-type Semiconductors Positively charged materials that enables them to carry current. They are produced by adding an impurity with three electrons in the outer ring (trivalent atoms).

Quick Release Valve A device used to exhaust air as close as possible to the service chambers or spring brakes.

Radial A tire design having cord materials running in a direction from the center point of the tire, usually from bead to bead.

Radial Load A load that is applied at 90° to an axis of rotation.

RAM An abbreviation for random access memory.

Ram Air Air that is forced into the engine or passenger compartment by the forward motion of the vehicle.

Random Access Memory (RAM) The memory used during computer operation to store temporary information. The microcomputer can write, read, and erase information from RAM in any order, which is why it is called random.

Range Shift Cylinder A component located in the auxiliary section of the transmission. This component, when directed by air pressure via low and high ports, shifts between high and low range of gears.

Range Shift Lever A lever located on the shift knob allows the driver to select low or high gear range.

Rated Capacity The maximum, recommended safe load that can be sustained by a component or an assembly without permanent damage.

Ratio Valve A valve used on the front or steering axle of a heavy-duty truck to limit the brake application pressure to the actuators during normal service braking.

RCRA An abbreviation for Resource Conservation and Recovery Act.

Reactivity The characteristic of a material that enables it to react violently with air, heat, water, or other materials.

Read Only Memory (ROM) A type of memory used in microcomputers to store information permanently.

Rear Hanger A bracket for mounting the rear of a truck or trailer suspension to the frame. Made to accommodate the end of the spring on spring suspensions. There are usually four types: flange-mount, straddle-mount, under-mount, and side-mount.

Recall Bulletin A bulletin that pertains to special situations that involve service work or replacement of parts in connection with a recall notice.

Reference Voltage The voltage supplied to a sensor by the computer, which acts as a base line voltage; modified by the sensor to act as an input signal.

Relay An electric switch that allows a small current to control a much larger one. It consists of a control circuit and a power circuit.

Relay/Quick Release Valve A valve used on trucks with a wheel base 254 inches (6.45 meters) or longer. It is attached to an air tank to main supply line to speed the application and release of air to the service chambers. It is similar to a remote control foot valve.

Refrigerant A liquid capable of vaporizing at a low temperature.

Refrigerant Management Center Equipment designed to recover, recycle, and recharge an air-conditioning system.

Release Bearing A unit within the clutch consisting of bearings that mount on the transmission input shaft but do not rotate with it.

Reserve Capacity Rating The ability of a battery to sustain a minimum vehicle electrical load in the event of a charging system failure.

Resistance The opposition to current flow in an electrical circuit.

Resisting Bending Moment A measurement of frame rail strength derived by multiplying the section modulus of the rail by the yield strength of the material. This term is universally used in evaluating frame rail strength.

Resource Conservation and Recovery Act (RCRA) A law that states that after using a hazardous material, it must be properly stored until an approved hazardous waste hauler arrives to take them to the disposal site.

Reverse Elliot Axle A solid-beam front axle on which the steering knuckles span the axle ends.

Revolutions per Minute (rpm) The number of complete turns a member makes in one minute.

Right to Know Law A law passed by the federal government and administered by the Occupational Safety and Health Administration (OSHA) that requires any company that uses or produces hazardous chemicals or substances to inform its employees, customers, and vendors of any potential hazards that may exist in the workplace as a result of using the products.

Rigid Disc A steel plate to which friction linings, or facings, are bonded or riveted.

Rigid Fifth Wheel A platform that is fixed rigidly to a frame. This fifth wheel has no articulation or oscillation. It is generally used in applications where the articulation is provided by other means, such as an articulating upper coupler of a frame-less dump.

Rigid Torque Arm A member used to retain axle alignment and, in some cases, to control axle torque. Normally, one adjustable and one rigid arm are used per axle so the axle can be aligned.

Ring Gear (1) The gear around the edge of a flywheel. (2) A large circular gear such as that found in a final drive assembly.

Rocker Beam A suspension device used to transfer and maintain equal load distribution between two or more axles of a suspension.

Roll Axis The theoretical line that joins the roll center of the front and rear axles.

Roller Clutch A clutch designed with a movable inner race, rollers, accordion (apply) springs, and outer race. Around the inside diameter of the outer race are several cam-shaped pockets. The clutch assembly rollers and accordion springs are located in these pockets.

Rollers A hardware part that attaches to the web of the brake shoes by means of roller retainers. The rollers, in turn, ride on the end of an S-cam.

ROM An abbreviation for read only memory.

Rotary Oil Flow A condition caused by the centrifugal force applied to the fluid as the converter rotates around its axis.

Rotation A term used to describe the fact that a gear, shaft, or other device is turning.

rpm An abbreviation for revolutions per minute.

Rotor (1) A part of the alternator that provides the magnetic fields necessary to create a current flow. (2) The rotating member of an assembly.

Runout A deviation of the specified normal travel of an object. The amount of deviation or wobble a shaft or wheel has as it rotates. Runout is measured with a dial indicator.

Safety Factor (SF) (1) The amount of load which can safely be absorbed by and through the vehicle chassis frame members. (2) The difference between the stated and rated limits of a product, such as a grinding disk.

Screw Pitch Gauge A gauge used to provide a quick and accurate method of checking the threads per inch of a nut or bolt.

Secondary Lock The component or components of a fifth wheel locking mechanism that can be included as a backup system for the primary locks. The secondary lock is not required for the fifth wheel to function and can be either manually or automatically applied. On some designs, the engagement of the secondary lock can only be accomplished if the primary lock is properly engaged.

Section Height The tread center to bead plane on a tire.

Section Width The measurement on a tire from sidewall to sidewall.

Self-Adjusting Clutch A clutch that automatically takes up the slack between the pressure plate and clutch disc as wear occurs.

Semiconductor A solid state device that can function as either a conductor or an insulator, depending on how its structure is arranged.

Semifloating Axle An axle type whereby drive power from the differential is taken by each axle half-shaft and transferred directly to the wheels. A single bearing assembly, located at the outer end of the axle, is used to support the axle half-shaft.

Semioscillating A term that generally describes a fifth wheel type that oscillates or articulates about an axis perpendicular to the vehicle centerline.

Semitrailer A load-carrying vehicle equipped with one or more axles and constructed so that its front end is supported on the fifth wheel of the truck tractor that pulls it.

Sensing Voltage The voltage that allows the regulator to sense and monitor the battery voltage level.

Sensor An electronic device used to monitor relative conditions for computer control requirements.

Series Circuit A circuit that consists of two or more resistors connected to a voltage source with only one path for the electrons to follow.

Series/Parallel Circuit A circuit designed so that both series and parallel combinations exist within the same circuit.

Service Bulletin A publication that provides the latest service tips, field repairs, product improvements, and related information of benefit to service personnel.

Service Manual A manual, published by the manufacturer, that contains service and repair information for all vehicle systems and components.

Shift Bar Housing Available in standard- and forward-position configurations, a component that houses the shift rails, shift yokes, detent balls and springs, inter-lock balls, and pin and neutral shaft.

Shift Fork The Y-shaped component located between the gears on the main shaft that, when actuated, cause the gears to engage or disengage via the sliding clutches. Shift forks are located between low and reverse, first and second, and third and fourth gears.

Shift Rail Shift rails guide the shift forks using a series of grooves, tension balls, and springs to hold the shift forks in gear. The grooves in the forks allow them to interlock the rails, and the transmission cannot be accidentally shifted into two gears at the same time.

Shift Tower The main interface between the driver and the transmission, consisting of a gearshift lever, pivot pin, spring, boot and housing.

Shift Yoke A Y-shaped component located between the gears on the main shaft that, when actuated, cause the gears to engage or disengage via the sliding clutches. Shift yokes are located between low and reverse, first and second, and third and fourth gears.

Shock Absorber A hydraulic device used to dampen vehicle spring oscillations for controlling body sway and wheel bounce, and/or prevent spring breakage.

Short Circuit An undesirable connection between two worn or damaged wires. The short occurs when the insulation is worn between two adjacent wires and the metal in each wire contacts the other, or when the wires are damaged or pinched.

Single-Axle Suspension A suspension with one axle.

Single Reduction Axle Any axle assembly that employs only one gear reduction through its differential carrier assembly.

Slave Valve A valve to help protect gears and components in the transmission's auxiliary section by permitting range shifts to occur only when the transmission's main gearbox is in neutral. Air pressure from a regulator signals the slave valve into operation.

Slide Travel The distance that a sliding fifth wheel is designed to move.

Sliding Fifth Wheel A specialized fifth wheel design that incorporates provisions to readily relocate the kingpin center forward and rearward, which affects the weight distribution on the tractor axles and/or overall length of the tractor and trailer.

Slipout A condition that generally occurs when pulling with full power or decelerating with the load pushing. Tapered or worn clutching teeth will try to "walk" apart as the gears rotate, causing the sliding clutch and gear to slip out of engagement.

Slip Rings and Brushes Components of an alternator that conducts current to the rotor. Most alternators have two slip rings mounted directly on the rotor shaft; they are insulated from the shaft and from each other. A spring loaded carbon brush is located on each slip ring to carry the current to and from the rotor windings.

Solenoid An electromagnet that is used to perform work, made with one or two coil windings wound around an iron tube.

Solid-State Device A device that requires very little power to operate, is very reliable, and generates very little heat.

Solid Wires A single-strand conductor.

Solvent A substance which dissolves other substances.

Spade Fuse A term used for blade fuse.

Spalling Surface fatigue occurs when chips, scales, or flakes of metal break off due to fatigue rather than wear. Spalling is usually found on splines and U-joint bearings.

Specialty Service Shop A shop that specializes in areas such as engine rebuilding, transmission/axle overhauling, brake, air conditioning/heating repairs, and electrical/electronic work.

Specific Gravity The scientific measurement of a liquid based on the ratio of the liquid's mass to an equal volume of distilled water.

Spiral Bevel Gear A gear arrangement that has a drive pinion gear that meshes with the ring gear at the centerline axis of the ring gear. This gearing provides strength and allows for quiet operation.

Splined Yoke A yoke that allows the drive shaft to increase in length to accommodate movements of the drive axles.

Spontaneous Combustion A process by which a combustible material ignites by itself and starts a fire.

Spread Tandem Suspension A two-axle assembly in which the axles are spaced to allow maximum axle loads under existing regulations. The distance is usually more than 55 inches.

Spring A device used to reduce road shocks and transfer loads through suspension components to the frame of the trailer. There are usually four basic types: multileaf, monoleaf, taper, and air springs.

Spring Chair A suspension component used to support and locate the spring on an axle.

Spring Deflection The depression of a trailer suspension when the springs are placed under load.

Spring Rate The load required to deflect the spring a given distance, (usually one inch).

Spring Spacer A riser block often used on top of the spring seat to obtain increased mounting height.

Stability A relative measure of the handling characteristics which provide the desired and safe operation of the vehicle during various maneuvers.

Stabilizer A device used to stabilize a vehicle during turns; sometimes referred to as a sway bar.

Stabilizer Bar A bar that connects the two sides of a suspension so that cornering forces on one wheel are shaped by the other. This helps equalize wheel side loading and reduces the tendency of the vehicle body to roll outward in a turn.

Staff Test A test performed when there is an obvious malfunction in the vehicle's power package (engine and transmission), to determine which of the components is at fault.

Stand Pipe A type of check valve which prevents reverse flow of the hot liquid lubricant generated during operation. When the universal joint is at rest, one or more of the cross ends will be up. Without the stand pipe, lubricant would flow out of the upper passage ways and trunnions, leading to partially dry startup.

Starter Circuit The circuit that carries the high current flow within the system and supplies power for the actual engine cranking.

Starter Motor The device that converts the electrical energy from the battery into mechanical energy for cranking the engine.

Starting Safety Switch A switch that prevents vehicles with automatic transmissions from being started in gear.

Static Balance Balance at rest, or still balance. It is the equal distribution of the weight of the wheel and tire around the axis of rotation so that the wheel assembly has no tendency to rotate by itself regardless of its position.

Stationary Fifth Wheel A fifth wheel whose location on the tractor frame is fixed once it is installed.

Stator A component located between the pump/impeller and turbine to redirect the oil flow from the turbine back into the impeller in the direction of impeller rotation with minimal loss of speed or force.

Stator Assembly The reaction member or torque multiplier supported on a free wheel roller race that is splined to the valve and front support assembly.

Steering Gear A gear set mounted in a housing that is fastened to the lower end of the steering column used to multiply driver turning force and change rotary motion into longitudinal motion.

Steering Stabilizer A shock absorber attached to the steering components to cushion road shock in the steering system, improving driver control in rough terrain and protecting the system.

Stepped Resistor A resistor designed to have two or more fixed values, available by connecting wires to either of the several taps.

Still Balance Balance at rest; the equal distribution of the weight of the wheel and tire around the axis of rotation so that the wheel assembly has no tendency to rotate by itself regardless of its position.

Stoplight Switch A pneumatic switch that actuates the brake lights. There are two types: (1) A service stoplight switch that is located in the service circuit, actuated when the service brakes are applied. (2) An emergency stoplight switch located in the emergency circuit and actuated when a pressure loss occurs.

Storage Battery A battery to provide a source of direct current electricity for both the electrical and electronic systems.

Stranded Wire Wire that is are made up of a number of small solid wires, generally twisted together, to form a single conductor.

Structural Member A primary load-bearing portion of the body structure that affects its over-the-road performance or crash-worthiness.

Sulfation A condition that occurs when sulfate is allowed to remain in the battery plates for a long time, causing two problems: (1) It lowers the specific gravity levels, increasing the danger of freezing at low temperatures. (2) In cold weather a sulfated battery may not have the reserve power needed to crank the engine.

Suspension A system whereby the axle or axles of a unit are attached to the vehicle frame, designed in such a manner that road shocks from the axles are dampened through springs reducing the forces entering the frame.

Suspension Height The distance from a specified point on a vehicle to the road surface when not at curb weight.

Swage To reduce or taper.

Sway Bar A component that connects the two sides of a suspension so that cornering forces on one wheel are shared by the other. This helps equalize wheel side loading and reduces the tendency of the vehicle body to roll outward in a turn.

Switch A device used to control on/off and direct the flow of current in a circuit. A switch can be under the control of the driver or can be self-operating through a condition of the circuit, the vehicle, or the environment.

Synchromesh A mechanism that equalizes the speed of the gears that are clutched together.

Synchro-transmission A transmission with mechanisms for synchronizing the gear speeds so that the gears can be shifted without clashing, thus eliminating the need for double-clutching.

System Protection Valve A valve to protect the brake system against an accidental loss of air pressure, buildup of excess pressure, or backflow and reverse air flow.

Tachometer An instrument that indicates rotating speeds, sometimes used to indicate crankshaft rpm.

Tag Axle The rearmost axle of a tandem axle tractor used to increase the load-carrying capacity of the vehicle.

Tapped Resistor A resistor designed to have two or more fixed values, available by connecting wires to either of the several taps.

Tandem One directly in front of the other and working together.

Tandem Axle Suspension A suspension system consisting of two axles with a means for equalizing weight between them.

Tandem Drive A two-axle drive combination.

Tandem Drive Axle A type of axle that combines two single axle assemblies through the use of an interaxle differential or power divider and a short shaft that connects the two axles together.

Three-Speed Differential A type of axle in a tandem two-speed axle arrangement with the capability of operating the two drive axles in different speed ranges at the same time. The third speed is actually an intermediate speed between the high and low range.

Throw (1) The offset of a crankshaft. (2) The number of output circuits of a switch.

Tie-Rod Assembly A system that transfers the steering motion to the opposite, passenger side steering knuckle. It links the two steering knuckles together and forces them to act in unison.

Time Guide Prepared reference material used for computing compensation payable by the truck manufacturer for repairs or service work to vehicles under warranty, or for other special conditions authorized by the company.

Timing (1) A procedure of marking the appropriate teeth of a gear set prior to installation and placing them in proper mesh while in the transmission. (2) The combustion spark delivery in relation to the piston position.

Toe A suspension dimension that reflects the difference in the distance between the extreme front and rear of the tire.

Toe In A suspension dimension whereby the front of the tire points inward toward the vehicle.

Toe Out A suspension dimension whereby the front of the tire points outward from the vehicle.

Top U-Bolt Plate A plate located on the top of the spring and is held in place when the U-bolts are tightened to clamp the spring and axle together.

Torque To tighten a fastener to a specific degree of tightness, generally in a given order or pattern if multiple fasteners are involved on a single component.

Torque and Twist A term that generally refers to the forces developed in the trailer and/or tractor frame that are transmitted through the fifth wheel when a rigid trailer, such as a tanker, is required to negotiate bumps, like street curbs.

Torque Converter A component device, similar to a fluid coupling, that transfers engine torque to the transmission input shaft and can multiply engine torque by having one or more stators between the members.

Torque Limiting Clutch Brake A system designed to slip when loads of 20 to 25 pound–feet (27 to 34 N) are reached protecting the brake from overloading and the resulting high heat damage.

Torque Rod Shim A thin wedge-like insert that rotates the axle pinion to change the U-joint operating angle.

Torsional Rigidity A component's ability to remain rigid when subjected to twisting forces.

Torsion Bar Suspension A type of suspension system that utilizes torsion bars in lieu of steel leaf springs or coil springs. The typical torsion bar suspension consists of a torsion bar, front crank, and rear crank with associated brackets, a shackle pin, and assorted bushings and seals.

Total Pedal Travel The complete distance the clutch or brake pedal must move.

Toxicity A statement of how poisonous a substance is.

Tracking The travel of the rear wheels in a parallel path with the front wheels.

Tractor A motor vehicle, without a body, that has a fifth wheel and is used for pulling a semitrailer.

Tractor Protection Valve A device that automatically seals off the tractor air supply from the trailer air supply when the tractor system pressure drops to 30 or 40 psi (207 to 276 kPa).

Tractor/Trailer Lift Suspension A single axle air ride suspension with lift capabilities commonly used with steerable axles for pusher and tag applications.

Trailer A platform or container on wheels pulled by a car, truck, or tractor.

Trailer Hand Control Valve A device located on the dash or steering column and used to apply only the trailer brakes; primarily used in jackknife situations.

Trailer Slider A movable trailer suspension frame that is capable of changing trailer wheelbase by sliding and locking into different positions.

Transfer Case An additional gearbox located between the main transmission and the rear axle to transfer power from the transmission to the front and rear driving axles.

Transistor An electronic device produced by joining three sections of semiconductor materials. Like the diode, it is very useful as a switching device, functioning as either a conductor or an insulator.

Transmission A device used to transmit torque at various ratios and that can usually also change the direction of the force of rotation.

Transverse Vibrations A condition caused by an unbalanced driveline or bending movements, in the drive shaft.

Treadle A dual brake valve that releases air from the service reservoirs to the service lines and brake chambers. The valve includes a piston which pushes on diaphragms to open ports; these vent air to service lines in the primary and secondary systems.

Treadle Valve A foot-operated brake valve that controls air pressure to the service chambers.

Tree Diagnosis Chart A chart used to provide a logical sequence for what should be inspected or tested when troubleshooting a repair problem.

Triaxle Suspension A suspension consisting of three axles with a means of equalizing weight between axles.

Trunnion The end of the universal cross; they are case hardened ground surfaces on which the needle bearings ride.

TTMA An abbreviation for Truck and Trailer Manufacturers Association.

Turbine The output (driven) member that is splined to the forward clutch of the transmission and to the turbine shaft assembly.

TVW An abbreviation for (1) Total vehicle weight. (2) Towed vehicle weight.

Two-Speed Axle Assembly An axle assembly having two different output ratios from the differential. The driver selects the ratios from the controls located in the cab of the truck.

U-Bolt A fastener used to clamp the top U-bolt plate, spring, axle, and bottom U-bolt plate together. Inverted (nuts down) U-bolts cross springs when in place; conventional (nuts up) U-bolts wrap around the axle.

UNEP An abbreviation for United Nations Environment Program. Mandates the complete phaseout of CFC-based refrigerants by 1995.

Underslung Suspension A suspension in which the spring is positioned under the axle.

United Nations Environmental Program (UNEP) A protocol that mandated the complete phase-out of CFC-based refrigerants by the year 1995.

Universal Gladhand A term often used for non-polarized gladhand.

Universal Joint (U-joint) A component that allows torque to be transmitted to components that are operating at different angles.

Upper Coupler The flat load-bearing surface under the front of a semitrailer, including the kingpin, which rests firmly on the fifth wheel when coupled.

Vacuum Air below atmospheric pressure. There are three types of vacuums important to engine and component function: manifold vacuum, ported vacuum, and venturi vacuum. The strength of either of these vacuums depend on throttle opening, engine speed, and load.

Validity List A list supplied by the manufacturer of valid bulletins.

Valve Body and Governor Test Stand A specialized piece of test equipment. The valve body of the transmission is removed from the vehicle and mounted into the test stand. The test stand duplicates all vehicle running conditions, so the valve body can be thoroughly tested and calibrated.

Variable Pitch Stator A stator design often used in torque converters in off-highway applications such as dirt and stone aggregate dump or haul trucks, or other specialized equipment used to transport unusually heavy loads in rough terrain.

Vehicle Body Clearance (VBC) The distance from the inside of the inner tire to the spring or other body structures.

Vehicle On-board Radar (VORAD) A system similar to an electronic eye that constantly monitors other vehicles on the road to give the driver additional reaction time to respond to potential dangers.

Vehicle Retarder An optional type of braking device that has been developed and successfully used over the years to supplement or assist the service brakes on heavy-duty trucks.

Vertical Load Capacity The maximum, recommended vertical downward force that can be safely applied to a coupling device.

VIN An abbreviation for Vehicle Identification Number.

Viscosity The ability of an oil to maintain proper lubricating quality under various conditions of operating speed, temperature, and pressure. Viscosity describes oil thickness or resistance to flow.

Volt The unit of electromotive force.

Voltage Generating Sensors These are devices which produces their own input voltage signal.

Voltage Limiter A device that provides protection by limiting voltage to the instrument panel gauges to approximately 5 volts.

Voltage Regulator A device that controls the amount of current produced by the alternator or generator and thus the voltage level in the charging circuit.

VORAD An acronym for Vehicle On-board Radar.

Vortex Oil Flow The circular flow that occurs as the oil is forced from the impeller to the turbine and then back to the impeller.

Watt The measure of electrical power.

Watt's Law A basic law of electricity used to find the power of an electrical circuit expressed in watts. It states that power equals the voltage multiplied by the current, in amperes.

Wear Compensator A device mounted in the clutch cover having an actuator arm that fits into a hole in the release sleeve retainer.

Wedge-Actuated Brakes A brake system using air pressure and air brake chambers to push a wedge and roller assembly into an actuator that is located between adjusting and anchor pistons.

Wet Tank A supply reservoir.

Wheel Alignment The mechanics of keeping all the parts of the steering system in the specified relation to each other.

Wheel and Axle Speed Sensors Electromagnetic devices used to monitor vehicle speed information for an antilock controller.

Wheel Balance The equal distribution of weight in a wheel with the tire mounted. It is an important factor which affects tire wear and vehicle control.

Windings (1) The three separate bundles in which wires are grouped in the stator. (2) The coil of wire found in a relay or other similar device. (3) That part of an electrical clutch that provides a magnetic field.

Work (1) Forcing a current through a resistance. (2) The product of a force.

Yield Strength The highest stress a material can stand without permanent deformation or damage, expressed in pounds per square inch (psi).

Yoke Sleeve Kit This can be installed instead of completely replacing the yoke. The sleeve is of heavy walled construction with a hardened steel surface having an outside diameter that is the same as the original yoke diameter.

Zener Diode A variation of the diode, this device functions like a standard diode until a certain voltage is reached. When the voltage level reaches this point, the zener diode will allow current to flow in the reverse direction. Zener diodes are often used in electronic voltage regulators.